D0907166

PRACTICAL ASPECTS OF GAS CHROMATOGRAPHY/ MASS SPECTROMETRY

PRACTICAL ASPECTS OF GAS CHROMATOGRAPHY/ MASS SPECTROMETRY

GORDON M. MESSAGE

A Wiley-Interscience Publication

JOHN WILEY & SONS

New York   Chichester   Brisbane   Toronto   Singapore

Library of Congress Cataloging in Publication Data:

Message, Gordon M., 1950–
    Practical aspects of gas chromatography/mass spectrometry.

    "A Wiley Interscience publication."
    Bibliography: p.
    Includes index.
    1. Gas chromatography.   2. Mass spectrometry.
I. Title
QD79.C45M46   1984   543′.0896   83-23475
ISBN 0-471-06277-4

Printed in the United States of America

10 9 8 7 6 5 4 3 2 1

Gas chromatography is known as one of the separation sciences although the word chromatography literally means colored writing.

Mass spectrometry also implies separation, separation of the masses into different groups.

And yet the practice of GC/MS transcends boundaries of class or color. My experience in traveling the world visiting GC/MS users has been one of fellowship. This book is dedicated to those people who apply themselves and their instrumentation to make the world a better and safer place to live in.

PREFACE

The GC/MS instrument is a complex system and is becoming even more complex with the increasing use of computer control and data-handling facilities. Operation of the instrument requires a basic knowledge of a number of scientific disciplines that are often outside the experience and training of the instrument user. A greater understanding of the system will enable the average user to get better results with the instrument, and it is also likely to increase his or her chances of getting valid results on the first run. This latter point is of some significance where the available sample is very limited.

The use of the word *system* is important. Expertise in chromatography and mass spectrometry separately does not guarantee success with GC/MS. The instrument is a true combination of its separate parts. Each half has to work with the other and imposes constraints on the overall system design. It is also important to realize that once you have bought your instrument you are stuck with its features. Few people have the means, let alone the inclination, to design or redesign their equipment. Hence, a compromise must be reached between the operational requirements of each part of the system, so that the overall performance is optimized.

GC/MS operators are often experts in their own field of study, and the GC/MS system is one of the tools at their disposal. Naturally enough, they want to get the best results from the instrument. This book is not intended as a replacement for the manufacturers' handbooks and manuals but rather as a bridge between the technicalities of the instrument and the science of GC/MS.

This book is by no means a learned text. I have included very few references. However, I have included some suggestions for further reading in the Bibliography. To a scholar, the key words are obvious, and a search in a good library will be fruitful. To those using GC/MS instruments or specifying GC/MS methods, here is some food for thought, based on my experiences in laboratories around the world. Be warned, it is full of opinion. In the end you should makeup your own mind about a technique or procedure, basing your decisions on facts and sound judgment. If I make you think about your methods, I will have succeeded. There is no single right way but certainly many wrong ways to operate a GC/MS system.

As I grew in experience at Finnigan Instruments, Ltd. I became more involved with customer training as well as routine service work. I was often struck by the lack of knowledge about the instrument's "works." Not that I wish to imply that there was any unwillingness to learn. On the contrary, whenever I visited a laboratory someone would ask: "What does this bit do?" Or, "How does that work?" In most cases a long and involved explanation was not required. The questioner did not want to design a GC/MS system. Quite often this person would not even be going to use it; he/she worked next door, upstairs, or wherever and was just curious. In part, this book is for those questioners.

Right from the beginning of my activities in field service and customer training work, I enjoyed working as part of a team with my customers, learning from and with them. The first two laboratories where I installed GC/MS systems were in my opinion above average. They set a standard for me, a way of working, that I have always tried to maintain and pass on to others. I should like to take this opportunity to thank Don Kirkwood, Robin Law, and their colleagues at the MAFF Laboratories, Burnham-on-Crouch, and Tim Webber and his colleagues at Shell Research, Sittingbourne, for their help and encouragement in my GC/MS salad days.

Many people have helped in the preparation of this manuscript. I would like particularly to acknowledge the help and encouragement of the following. Frank Davis, International Service Manager, Finnigan–MAT encouraged me to start, saying it was a very worthwhile and personally rewarding project. Bill McFadden, also of Finnigan, advised me not to start, saying it would be many years of hard slog, but on seeing my determination went out of his way to help. (They were both right!) My colleagues at Finnigan Ltd. have been a continuing source of encouragement, in particular, John Wellby of Shell Research, who later joined us as Application Manager. John also became my neighbor, and we have worked together and talked over many ideas both in and out of work as a result. I should also like to thank Barbara Wellby, John's wife, for carefully reading through my manuscript.

Two major groups deserve special mention. These are the GC/MS manufacturers and their customers. With my association with Finnigan it is natural that there should be a lot of their material in this book, but all the other manufacturers approached were both encouraging and helpful and I thank them for this. Many GC/MS users, the customers, contributed material and ideas. I have tried to give credit where it is due and if I have left anyone out, I apologize.

Finally, I should like to stress that this work was a private venture and has only been possible with the help of my family. Not only has my wife, Jane, given me continuous support but she also acted as my secretary, handling a great deal of the paperwork. My mother-in-law, Min, did all the manuscript typing and her husband, Ron, with a lifetime of journalistic experience, gave me a lot of helpful advice. Simon and Emma, my children, allowed me to share the study with them. Their interest and enthusiasm toward my work have always been a great source of encouragement.

GORDON M. MESSAGE

Leighton Buzzard, England
May 1984

CONTENTS

PART 3   TROUBLESHOOTING FAULTS ON GC/MS SYSTEMS

PART 4   CHOOSING A GC/MS SYSTEM

PRACTICAL ASPECTS
OF GAS
CHROMATOGRAPHY/
MASS SPECTROMETRY

CHAPTER ONE

An Introduction
to GC/MS Systems

The combination of the gas chromatograph with the mass spectrometer has resulted in an instrument of considerable importance in the field of chemical analysis and detection. Figure 1 shows two typical laboratory arrangements. It could be argued that the mass spectrometer is just one of a range of detectors available to the gas chromatograph user. On the other hand, some would say that the gas chromatograph is only one of many inlets that could be connected to the mass spectrometer. In truth, the synergy of this arrangement goes beyond the possibilities of the two separate instruments and almost warrants a new definition as an instrument in its own right. For this reason, and for convenience, the combined arrangement is referred to as a GC/MS system throughout this text.

The principal elements of a GC/MS system are shown in Figure 2. A computer system is included in the diagram, as well as its interface with the instrument operator. Depending on design, and certainly on manual systems, some or all of the computer control functions are undertaken by the operator directly. In any given system it is probably the mass spectrometer that sets the overriding design parameters.

An essential element in good mass spectrometer performance is a high-quality vacuum system. The pressure requirements vary with instrument type and application, but 1.3×10^{-3} Pa (10^{-5} torr) is a typical maximum, and often lower pressures are necessary. The gas chromatograph, by contrast, basically operates at or above atmospheric pressure, and so it is not surprising that some form of interface is required. This usually acts as a sample-enrichment device, extracting much of the carrier gas, although direct connection through tubing of a very fine bore is also possible. The latter arrangement simplifies interface design but places an increased burden on the mass spectrometer pumping system.

Interface design is discussed in detail in later chapters. All the enrichment

Figure 1. Typical GC/MS facilities. (*a*) A Finnigan 3200F with integral 6000 series data system. (Courtesy of Masspec Analytical Ltd.) (*b*) A MAT 212. The detachable GC is hiding in the trees to the left of the picture. (Finnigan-MAT, GmbH)

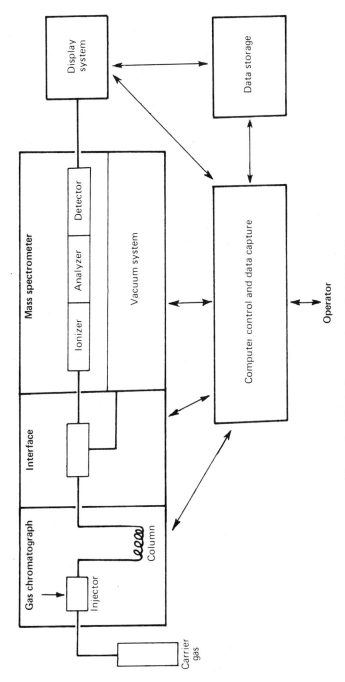

Figure 2. Block diagram of a typical GC/MS system.

types reflect the need to compromise between the amount of sample lost and the amount of carrier gas extracted. In general, they are more efficient with a light gas, and so helium is favored over nitrogen as the GC/MS carrier in most stand-alone gas chromatography work. As a result, many of the normal gas chromatograph parameters change when GC/MS systems are considered. Things are further complicated by the fact that the performance of the interface device is often dependent on the carrier gas flow rate and so it may be necessary to experiment with parameters to get the best from the system.

It is well to remember this last point. There may be a recommended or preferred value for a given parameter, but often significant improvements can be made by a deviation from the given value. In many applications it may not be worth the bother of experimentation, but occasionally it is vital to determine the limitations of the instrument parameters and the interactions between them.

While Part 1 (Chapters 2 to 6) concentrates on the theory of operation of each part of the GC/MS system, Part 2 (Chapters 7 to 10) reviews procedures and techniques in the light of this theory. Part 2 presents a number of ideas relating to instrument operation that are based on the author's experience of GC/MS laboratories around the world. It is important to realize that the methods and techniques presented here are not the only way to operate a GC/MS facility but are representative of the practices in successful laboratories. Applications are discussed primarily to illustrate topics rather than as the principal subject.

The increasing availability of sophisticated but low-cost computer systems has made possible a high level of instrument control and data handling. One might expect this to ease the GC/MS operator's burden. On the one hand, the data system can lead the user through the correct sequence of operations, even taking over some of the routine tasks. The computer can also provide a high level of system protection, self-checking, and control.

On the other hand, computer use opens up the way for complex control modes that were not practical before. Linked scans and automatic specific tuning are typical examples. Add to this the developments that have been made in data processing, such as precision mass assignment, data enhancement, and noise rejection, and the operator is faced with many new possibilities. Data outputs in the form of hard copies or video displays are also faster than could ever be achieved manually, assuming they could be achieved at all.

As a result it can be seen that computers make some routines possible and others easier. Unfortunately, it is all too easy to look at a beautifully drawn spectrum and believe it implicitly. The computer drew it, so it must be right! Sadly, this is not always the case, and once again it is those with a greater knowledge and understanding of their (even more complex) system that get the most reliable results. Hence Chapter 6 reviews the principles of computer operation that are applicable to GC/MS systems and throughout the text reference is made to data system methods as well as traditional manual approaches.

As instrument complexity increases, the likelihood of failure also increases. Failures can take many forms. They may be permanent and obvious, or they may be a nuisance but such that the instrument can still run. Worst of all, they may be intermittent. Part 2 ends with a chapter on preventive maintenance in the hope that failures can be minimized. It would be unrealistic to say that failures can be prevented, so Part 3 (Chapters 11 and 12) is devoted to troubleshooting and fault finding.

The GC/MS system is a powerful tool for chemical analysis that is capable of analyzing very small amounts of material. Typically, modern instruments can be used to identify compounds at about the 10^{-10} g level per microliter of solvent in the presence of other compounds. The system can also be set up to look for specific characteristic ions of known compounds or classes of compound. In this mode, detection limits can be as low as 10^{-12} g in quite complex mixtures, and lower levels under certain conditions are not unheard of. Unfortunately, not many GC/MS systems are operated and maintained ready to achieve the limit of their specifications.

This and other topics of special concern to readers responsible for specifying GC/MS methods and/or choosing a GC/MS system are discussed in Part 4 (Chapter 14).

—

PART ONE

GC/MS SYSTEMS AND COMPONENTS

CHAPTER TWO

Vacuum Systems

A large amount of the hardware visible on GC/MS instruments concerns the pumping and vacuum systems as shown in Figure 3. A very high vacuum is essential in the mass spectrometer for the following reasons:

1. High voltage breakdown may occur in the multiplier, source, or analyzer if the pressure becomes too great.
2. Oxygen from residual air and leaks will cause filament burnout in the ion source and ion gauges.
3. The mean free path of molecules in the system must be longer than the ion path through the analyzer, or ion-molecule collisions will occur.
4. As the pressure rises, regulation of the electron current through the ion source becomes more difficult.
5. Ion-molecular reactions start to occur as the pressure rises, producing changes in fragmentation patterns. This effect is used to advantage in chemical ionization (CI), but the pumping requirements for the analyzer relative to the ion source are even more stringent in the CI mode. Also, for good chemical ionization, the reagent gas pressure must be stable and controllable.
6. High background pressures are due to the presence of compounds in the mass spectrometer. These give interfering mass spectra, making interpretation difficult.
7. Contamination of ion-source components, slits or rods, and multipliers increases with higher pressures. This necessitates downtime for cleaning and affects running costs.

In addition to the very high vacuum system, some form of medium-to-high auxiliary vacuum arrangement is required for the following tasks:

1. Rough pumping of the mass spectrometer is necessary before the very high vacuum pumping system is applied, whenever the instrument has been vented to the atmosphere (for example, during source cleaning).

2. Rough pumping is needed to clear the reagent and carrier gas lines, to prevent cross contamination when new gases are selected.

3. The GC/MS interface usually contains a sample-enrichment device for electron impact work. This usually requires a pump to remove excess carrier gas.

4. Diverters are frequently used in the GC/MS interfaces to prevent the entry of solvent fronts and unwanted compounds of the GC eluent into the mass spectrometer. The unwanted carrier gas with solvent or sample may be pumped away by a bypass arrangement.

5. Auxiliary inlets, such as the probe and batch inlets, have interlocks that must be rough-pumped before they are connected to the main vacuum system.

In practice the auxiliary high vacuum system and the very high vacuum system are not entirely separate. Some pumps are used for several tasks, either together

(a)

Figure 3. Typical GC/MS vacuum systems. (a) A turbopump system. Note the spare backing pump connections for optional units. (Finnigan-MAT, GmbH)

(b)

Figure 3 (*continued*) (*b*) A diffusion pump system. (Nermag SA)

or at different times. In addition, there is an increase in the use of pneumatic or differential pressure controls for variable geometry/conductance sources and isolation valves. The pumping requirements for these tasks are modest, and the spare capacity of one of the system pumps is often used for this purpose.

VACUUM TERMS

Atmospheric Pressure. Varies with the weather, but for scientific purposes considered to be constant at 1.013×10^5 Pa (760 torr).

Conductance. In a pumping system, the inverse of pumping resistance. Conductance figures relate to the ease with which a vacuum system is pumped; the higher the conductance, the faster the pumping. Each element of a vacuum system has a conductance. The conductances add in an analogous way to conductivity (inverse of resistance) in electrical circuits. Generally speaking, the larger the diameter and straighter the path of the pumping system to the pumps, the greater is the system's conductance.

Differential Pumping.   The independent pumping of two regions of a vacuum system that are separated by a restriction.

Fore-Vacuum.   The same as rough vacuum; the vacuum usually provided by forepumps or roughing pumps.

High Vacuum (HV).   The pressure region at or below 10^{-3} torr.

Laminar or Molecular Flow.   The flow of a gas in which the motion of each gas molecule is considered to be independent of its neighbors and there are infrequent collisions.

Leaks and Virtual Leaks.   All vacuum systems leak, it is just a matter of degree. In practice we consider a leak to be any extra leakage that tends to increase the normal "leak-free" pressure. These leaks should be detectable by the mass spectrometer if a gas or solvent is applied to the leaking area, but in practice this is not always as simple a test as it seems. If gas or liquid is trapped inside the vacuum envelope, it may leak out into the system. For example, air trapped under a gasket seal within the mass spectrometer will give a "leak spectrum" but is not detectable from the outside. These are vitual leaks; they will eventually pump away, but this may take days or even weeks.

Mean Free Path.   Distance traveled by a molecule in its normal kinetic motion before it encounters another molecule.

Partial Pressure.   The pressure contribution due to each component of a mixture of gases.

Pascal.   The standard unit of pressure corresponding to a force of 1 newton per square meter (1 N/m^2); named after the French scientist Blaise Pascal.

Pumping Speed.   The amount of gas at a specified pressure pumped away by a pumping system expressed as volume per unit of time.

Rough Vacuum.   The vacuum usually provided by a rotary pump, typically in the range 1.3×10^2 to 1.3×10^{-1} Pa (1 to 10^{-3} torr). Sometimes referred to as *backing pressure* from the use of a rotary pump (or a forepump) to back a diffusion pump.

Torr.   A unit for measurement of pressure. Defined as the pressure required to support a column of mercury 1 mm high. Named after 17th-century vacuum researcher Evangelista Torricelli. 1 torr is equivalent to 133.332 Pa.

UltraHigh Vacuum (UHV).   The pressure region at or below 10^{-8} torr.

Vacuum.   Any pressure below atmospheric pressure.

Very High Vacuum (VHV).   Pressure at or below 10^{-6} torr.

Viscous or Turbulent Flow.   The flow of a gas in which the motion of each gas molecule is dependent on its neighbors and there are frequent collisions and mixing.

VACUUM PRESSURE UNITS AND TYPICAL SYSTEM PARAMETERS

The international standard of measure for pressure in the SI system of units is the pascal (Pa). Traditionally in high vacuum work the torr has been the base unit for measurement and it is still widely used. One torr is equivalent to 133.332 Pa, or 1 Pa is equivalent to 7.5×10^{-3} torr.

Typically, rotary pumps achieve pressures around 1 Pa, and diffusion pumps work in the millipascal range and lower. The pascal, defined as a force per unit area, is more meaningful at these low pressures than the torr, which is defined as the pressure needed to support a column of mercury 1 mm high. At 1.3×10^{-6} Pa (10^{-8} torr), the approximate limit for most GC/MS systems, the pressure is insufficient to support a column of mercury even one atom high. Nevertheless, even at this low pressure there will be about 3.8×10^{10} molecular impacts on each square millimeter of the vacuum chamber surface per second. Amazingly, the mean free path at this pressure is about 5 km, and the number of molecules per cubic centimeter is about 2.7×10^8.

At atmospheric pressure there are 2×10^{19} molecules per cubic centimeter. Each second, 3×10^{23} molecules "hit" each square centimeter of a container's surface, exerting a force on the container. The mean free path of each molecule is 6.5×10^{-8} m.*

For a typical mass spectrometer with a path length of 500 mm, the pressure must be reduced to 1.3×10^{-2} Pa (10^{-4} torr) or less throughout the mass spectrometer to prevent collisions. In practice, this means that the pressure must be 5 to 10 times lower in the main pumping region to allow for pumping speed restrictions caused by slits and lenses and for the statistical spread in actual free paths of all the molecules present. Further reductions in pressure are necessary to accommodate the high field strengths and very long path lengths encountered in some instruments.

*These figures are approximate only and are based on nitrogen gas at room temperature.

Hence, typical analyzer pressure limits are 2.7×10^{-3} Pa (2×10^{-5} torr) for quadrupole instruments and 2.7×10^{-4} Pa (2×10^{-6} torr) for magnetic instruments. There will, of course, be variations between different types of instruments and different limits for operation under different conditions. For example, analyzer pressures in differentially pumped chemical ionization instruments operating with a source reagent gas pressure of 66.7 Pa (0.5 torr) may rise to 6.7×10^{-3} Pa (5×10^{-5} torr) for quadrupole mass spectrometers and 6.7×10^{-4} Pa (5×10^{-6} torr) for magnetic instruments.

VACUUM COMPONENTS

The GC/MS vacuum system consists of a number of units and assemblies, as shown in Figure 4. The requirements will vary from one instrument to another, but the essential components are high-vacuum and low-vacuum pumps, pipe-

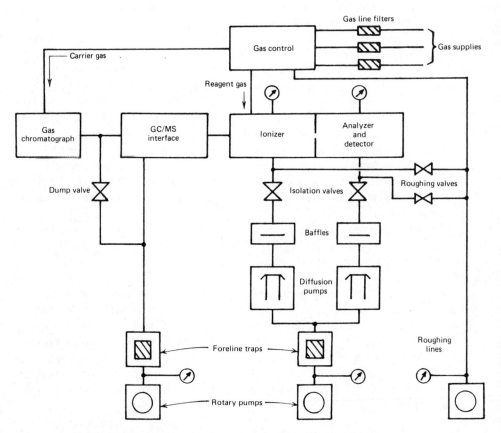

Figure 4. The basic GC/MS vacuum system arrangement.

work, baffles, valves, gauges, sensors, and gas supply controllers. In addition, there are unseen elements, such as pump oils and vacuum seals. Various aspects of these components and their relevance to GC/MS instruments are discussed in the following pages

Rotary Pumps

There are a number of different types of rotary pumps. All use the same principle of taking in a large volume of low-pressure gas and compressing it to a smaller volume by the action of a rotating device. The most common type used on GC/MS systems consists of a cylindrical rotor mounted eccentrically in a cylindrical chamber (Figure 5). Spring-loaded rotor blades are mounted diametrically through the rotor sealing against the inner surfaces of the chamber. The

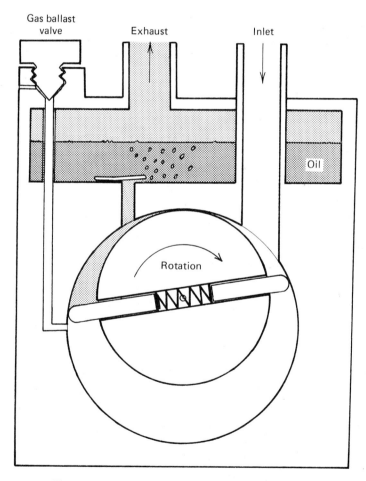

Figure 5.   Cross section of a rotary vacuum pump.

rotor is turned by an electric motor either directly or via a belt drive. The rotor blades are free to move in and out of the rotor and will sweep out a varying volume as the pump is turned. The gas inlet port is connected to the point of largest enclosed volume, and the outlet port is arranged to coincide with the smallest compressed volume. Gas is ejected through a spring-loaded flap valve, which prevents gas from reentering the pump.

The sealing is further improved by immersing the pumping assembly in an oil bath and allowing some of the oil into the pump chamber. This seals the small end gaps between the rotor blades and pumping chamber. The oil also serves as a lubricant. Small-bore oilways allow some oil along the drive shafts into the rotor chamber. This lubrication reduces pump wear by lowering friction between rotor blades and the cylinder walls. Even so, the pump will run quite hot. As a consequence, the oil must be suitable for high-temperature operation, producing a low vapor pressure even in the presence of air and other gases and vapors. Highly refined hydrocarbon oils are usually used, sometimes with additives to prevent oxidation.

The limiting pressure for a rotary pump depends principally on its compression ratio and the oil vapor pressure. To improve the compression ratio, pumps are often made with two chambers or stages connected in series. With good oils 1.3×10^{-2} Pa (10^{-4} torr) can be achieved, but this can be significantly degraded when pumping condensable vapors such as water vapor, as found in air, and solvents that are usually derived from divertors dumping the gas chromatograph solvent front. As the pumped gas is compressed, the partial pressure of an included vapor may reach its saturation vapor pressure. If this occurs before all the gas reaches atmospheric pressure, when the flap valve would normally open, the vapor will condense and will then be trapped as a liquid in the pump oil. When this oil reenters the pumping chamber via the oilways, the solvent or water may evaporate, thus setting a limit on the ultimate pressure obtainable.

Gas Ballasting.   To overcome the problem of condensation, a system of gas ballasting is used. Once the rotor passes the inlet port, the gas is trapped at a reduced pressure in a decreasing volume. If some air from outside is allowed to enter this space, the pressure quickly builds up, forcing open the flap valve before the vapor partial pressure reaches saturation. Condensation is therefore prevented. In addition, the large amount of gas pumped through the oil gives a large surface area for further vapor evaporation. The frequency with which a pump must be gas ballasted will depend on each application and the location in the pumping system. Some pumps may require ballasting once a week, while others may never need this attention. Pumping efficiency is reduced during ballasting, because air leakage past the rotor blade edges causes backstreaming into the vacuum system. Open ballast valves only a small amount, and monitor the vacuum pressures carefully to prevent vacuum accidents.

Even with ballasting, pump oils eventually become contaminated, and they should be changed at regular intervals if pumping speed is to be maintained. Pumping speed is a function of size and speed of rotation. A direct-drive pump usually has a greater pumping speed than a belt-driven unit of about the same physical size, but wear on the rotor blades and oil seals is greater.

Diffusion Pumps

The diffusion pump moves gas from one area to another by establishing conditions in which the mean free path for molecules in one direction is greater than in the opposite direction. The diffusion pump consists of an enclosed chamber in which a stack of concentric jets are mounted (Fig. 6). A small charge of oil covers the bottom of the chamber, which is heated strongly. The oil boils vigorously, producing a heavy vapor stream, which passes up the inside of the stack and out through the jets. These are angled downwards so that the escaping vapor produces a cloud of molecules moving rapidly toward the bottom of the pump. The outside of the pump is cooled, usually by a water jacket or cooling rings but sometimes by an air blower. The oil vapor is condensed and runs down to the bottom of the pump into the boiler to repeat the cycle.

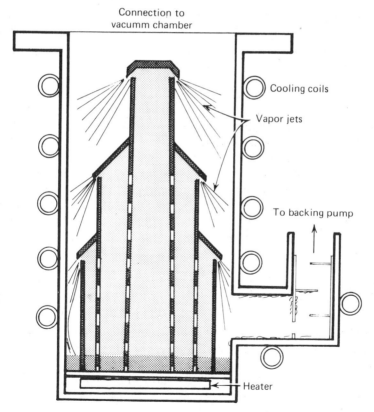

Figure 6. Cross section of a diffusion pump.

Any gas molecules that happen to diffuse into the pump will suddenly encounter a strong stream of particles moving predominantly downwards. Consequently, the gas molecules will be driven down to the lower part of the pump. A pressure difference of up to five orders of magnitude can be developed across the pump. Unfortunately, the diffusion pump cannot operate with its outlet at atmospheric pressure. Under such conditions there would be a number of problems:

The vapor jets would encounter such a large number of molecules that the flow would become turbulent and not necessarily downwards.

The oil would have to be heated much more strongly to achieve a sufficiently high boiling rate.

There would be significant oxidation of the oil and pump elements.

Even five orders of magnitude reduction in pressure would give only about 1.3 Pa (10^{-2} torr) inlet pressure if the pump operated with its outlet at atmospheric pressure.

Hence, it is necessary for the diffusion pump to be backed by another pump, which must be capable of operating with its inlet in the 1.3 to 0.13 Pa (10^{-2} to 10^{-3} torr) region and its outlet at atmospheric pressure. This is usually accomplished with a rotary vacuum pump.

The diffusion pump can create an ultimate vacuum of about 1.3×10^{-6} Pa (10^{-8} torr) with an outlet pressure of 0.13 Pa (10^{-3} torr). This may take some time to achieve depending on the pumping speed and the rate of outgassing of the system. One problem that can occur is that gas molecules entering the pump are condensed with the oil and are then boiled off, cycling around the pump and reducing its efficiency.

To overcome this problem the pump is designed so that a significant portion of the boiling oil, with impurities, is allowed to pump along the backing arm. A series of baffles cool the vapor slowly so that most of the oil condenses and runs back but the impurity is held in a warm gaseous state and is pumped away. There is always a steady but small loss of diffusion pump oil to the backing pump. Also, while the flow of oil vapor from the jets is predominantly downwards, some does escape upwards. The oil loss is usually so slow that the pumps need only be checked every 12 to 18 months.

Diffusion Pump Fluids

The ultimate pressure obtainable with a diffusion pump depends on the vapor pressure of the pumping fluid. Cold baffles reduce the ultimate pressure, but it is still the characteristics of the fluid that set the limit. Fluids of very low vapor pressure have to be used. The choice of fluid will depend on the pumping requirements in each application. The principal types available are discussed below.

Polyphenyl Ether Oils.   Organic oils used extensively in GC/MS systems. There are a number of types, and they should not be mixed. They are very susceptible to oxidation if operated with a large air throughput.

Silicone Oils.   Oils based on silicon polymers. They will tolerate more air than polyphenyl ether oils and so are often used in inlet vacuum interlocks. Unfortunately, they crack and eventually oxidize to form silica (silicon dioxide), a very hard insulating material. They are not suitable for areas where the critical instrument components cannot be physically cleaned. That is, they may be acceptable in source and slit areas that can be abrasively cleaned but are disastrous if deposited on electron multipliers. Hence, they are not generally used in diffusion pumps in the main system.

Mercury.   An efficient pump fluid, but it has a high vapor pressure, 0.13 Pa (10^{-3} torr), at room temperature. Some form of cold baffle is therefore essential, and there is a real danger of the mercury migrating from the pump to the baffle surface. Mercury vapor is also very corrosive and extremely toxic, and hence it is not often used in GC/MS systems.

Turbomolecular Pumps

The pumping action of a turbomolecular pump is similar in some respects to that of a diffusion pump, in that the mean free path of gas molecules is greater in one preferred direction. With diffusion pumps, this is due to collisions with molecules of the pump oil vapor. In turbomolecular pumps, the gas molecules collide with moving rotor blades. The mean free path in an open enclosure would be equal in all directions, but the collisions with the rotor impart energy to the gas in a preferred direction. Consequently the gas is moved, that is pumped, through the rotor/stator assembly. As the gas is pushed through the pump, its pressure is increased. If the pressure is increased to a point where the mean free path becomes small and a collision with another gas molecule rather than the rotor is more likely, then the pumping action will cease. To prevent this, the turbomolecular pump must be backed by another type of pump, usually a rotary mechanical pump.

A cross section of a typical turbomolecular pump is shown in Figure 7. A normal rotor/stator arrangement has a number of disk pairs, which act as pumps in series. As the pressure is increased along the rotor, the pumping characteristics of the rotor blades must vary. The blades near the high-vacuum end of the pump must have a high pumping speed, and compression ratio is sacrificed. At the low-vacuum, high-pressure end the compression ratio must be high, and pumping speed is compromised. The two types of blades can be seen in the photograph of a typical turbomolecular pump rotor in Figure 8. Rotor speeds vary with type, but a typical unit will operate at 40,000 rpm and some go as fast as 80,000 rpm. Consequently, the assembly must be very carefully balanced to prevent vibration.

At these high speeds the rotor bearings must be lubricated. The bearing oil

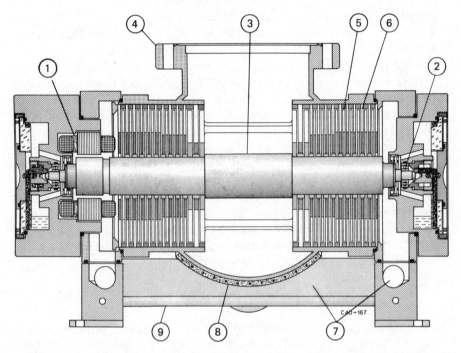

Figure 7. Sectional view of a Pfeiffer turbomolecular pump. *1*, Motor; *2*, oil-cooled bearing; *3*, rotor assembly; *4*, ultrahigh vacuum flange; *5*, rotor disk; *6*, stator disk; *7*, roughing vacuum line; *8*, heater; *9*, water cooler. (Arthur Pfeiffer GmbH)

could get quite hot and vaporize, streaming into the vacuum system if water or blown air cooling were not used. Even so, bearing life is limited and bearings and their oil have to be replaced at regular intervals. Changing bearings on single-rotor vertically mounted pumps is a job for a specialist, as the rotor has to be removed and balanced when it is replaced. On the horizontally mounted double-rotor versions, bearing replacement is usually more straightforward and can be done on site.

There is always the possibility of small particles, ionizer screws, or solids probe cups entering the GC/MS vacuum system. Even quite a small object hitting a rotor traveling at 50,000 rpm can cause it to disintegrate. To protect it, a fine mesh is usually fitted across the throat of the pump.

Turbomolecular pumps are usually vented directly to atmospheric pressure. Electronic trips are fitted to protect the drive motor and its controller from the overload. The pump is usually fitted with a venting flange. Air entering at this point will rush through the pump toward the backing pump lines. This prevents bearing oil from the rotary and turbomolecular pumps from being blown into the vacuum chamber. It is normal to fit a filter on the vent inlet to prevent particles from entering the pump, and often GC/MS manufacturers fit an air-drying

Figure 8. Rotor assembly of a pfeiffer TPU 510 turbomolecular pump. (Arthur Pfeiffer GmbH)

unit. This not only acts as a particle filter but also prevents moisture from entering the vacuum chamber, improving its subsequent pumpdown characteristics. Venting takes only about a minute; the pump stops quite quickly due to the friction of the air loading. Pumpdown takes longer, 10 min being typical.

Oil and Vapor Traps

Cold Baffles.   To prevent backstreaming of oil out of the top of diffusion pumps, some form of high conductance cold baffle is often used. Baffles can take the form of chevrons, disks, rings, or concentric cylinders and are usually cooled. With concentric cylinders, the inner cylinder has a tube to the outside so that it can be filled with refrigerated alcohol or liquid nitrogen. The other types can be cooled with liquids at either normal or reduced temperatures or by means of thermoelectric devices known as Peltier effect coolers.

Apart from preventing oil loss from the diffusion pump, the cold baffles can act as pumps. If molecules coming into contact with the surface are cooled sufficiently they will condense and no longer contribute to the partial pressure of the gas in the vacuum system. Other molecules hitting the baffle may cause them to be knocked off with sufficient energy to reenter the gas phase. Hence the cold baffles are efficient pumps only when a very high vacuum has already been established.

Foreline Traps.   There is always a small amount of oil vapor backstreaming from a rotary pump. This backstreaming will increase when the pumps have a significant gas throughput or when there is a leak in the pumping lines. The oil mist entering the pumped enclosure is undesirable. It gives rise to a high background spectrum in the mass spectrometer and reduces the efficiency of diffusion pumps. Source and transfer line contamination is also a problem when backstreaming occurs in interface backing pumps. To prevent this contamination, a trap can be fitted in the pumping line immediately above the rotary pump inlet port. These foreline traps consist of a relatively large chamber containing pellets of a molecular sieve material. The chamber is made large enough to give a satisfactory conductance. Oil mist and some vapors are absorbed by the molecular sieve and hence do not stream further back down the line.

In time the molecular sieve becomes saturated and loses its efficiency. In addition to oil vapor, it may well absorb significant amounts of water and solvent vapors. It is then necessary to recharge the filter trap. The old sieve pellets can be rejuvenated by baking. This is sometimes done in situ, by an enclosed foreline trap heater. It is then necessary to isolate the pumping line immediately above the trap to prevent massive backstreaming. To improve the pumping of the outgassed water and solvent, it is normal to open the gas ballast valves while the foreline traps are being heated. Alternatively, the system may be shut down and the molecular sieve material replaced.

Vacuum Gauges

Pirani Gauges.   Operating in the region from 1333 Pa (10 torr) down to 0.13 Pa $(10^{-3}$ torr), Pirani gauges are used to measure the pressure in backing pump lines and other areas normally evacuated by rotary pumps. The gauge contains resistance elements that form part of a resistance bridge or Wheatstone net circuit. The bridge out-of-balance voltage is a function of the vacuum pressure.

A typical arrangement is shown in Figure 9. The bridge circuit consists of two stable resistors R1 and R2 and the Pirani gauge resistors R3 and R4. A voltage is applied across the resistor arrangement. The Pirani gauge resistors are very temperature dependent. They are heated by the current flowing through them, and their resistance changes. Resistor R3 is cooled by the surrounding gas, and so its resistance is a function of the local pressure. Resistor R4 is not cooled by the surrounding gas in the vacuum system but by thermal contact with the gauge enclosure. As a result, it compensates for ambient temperature changes. The bridge out-of-balance voltage, V_B, is therefore a function of the pressure and independent of operating temperature changes. The greater the vacuum pressure, the greater the number of cooling gas molecules and hence the greater the out-of-balance voltage. This voltage could be displayed directly on a galvanometer calibrated in pressure units, but it is more usual to feed the signal via buffer amplifiers to linearizing and display circuits.

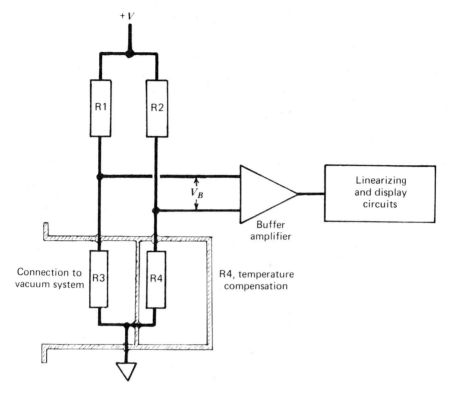

Figure 9.   Vacuum measurement using a Pirani gauge.

Thermocouple Gauge.   As with Pirani gauges, thermocouple gauges are used in backing lines; and they operate from about 1.3×10^4 Pa (100 torr) down to 1.3×10^{-1} Pa (10^{-3} torr). They are extremely simple in construction and require simple driving circuits, but their response characteristic is very nonlinear.

The gauge consists of two dissimilar wires mounted in the vacuum enclosure to form a cross. They are spot welded together at the center. One of the wires is heated by an applied ac voltage, as shown in Figure 10. The mains line voltage is converted to a symmetrical square wave by a limiting circuit and applied to a driving transformer. The output of the transformer is used to heat the gauge wire AC. Wire BD forms a thermocouple junction with wire AC. A metering circuit connected between the center-tapped transformer output and the thermocouple will sense the temperature-dependent voltage but not the heating ac voltage. A temperature-dependent resistor (thermistor) R1 is usually fitted inside the gauge connector to compensate for ambient temperature changes. The temperature of the wire is dependent on the applied voltage and on the rate of cooling by the surrounding gas. The heating voltage is made independent of mains variations by the limiting circuit, and so the output voltage of the thermocouple is a func-

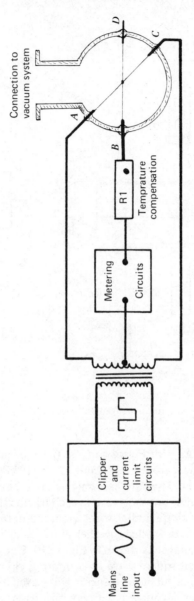

Figure 10. Vacuum measurement using a thermocouple gauge.

tion of the gauge pressure. Although normally provided, the output connection D is not needed in this circuit arrangement.

Ion Gauge. Below about 1.3×10^{-1} Pa (10^{-3} torr) there are insufficient gas molecules for cooling effects to be used as a gauge system, and ionization techniques must be used. The Bayard-Alpert ion gauge can measure pressures of 1.3×10^{-6} Pa (10^{-8} torr) or less, with an upper limit of about 1.3×10^{-1} Pa (10^{-3} torr). Electrons emitted from a heated filament (Fig. 11) are accelerated toward an open grid. They collide with gas molecules to produce positive ions by an electron-impact mechanism. The positive ions are collected, and the resultant current, which is a function of the ion gauge pressure, is measured.

A control circuit regulates the emission of electrons from the filament by controlling the filament-heating current and hence the temperature. Typically the filament is held at -30 V with respect to ground. It is mounted outside a grid, normally an open coil, which is held at about $+150$ V. Electrons are therefore accelerated toward the grid and may make several traverses before reaching it.

If an electron comes close to a gas molecule, a positive ion may be formed as a

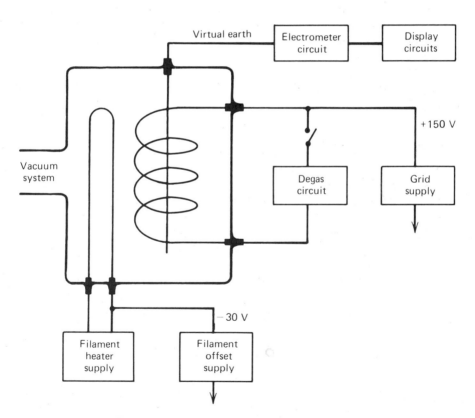

Figure 11. Typical ion gauge control circuits.

result of an inelastic collision. The number of ions formed is a function of the gas pressure. Ions formed inside the coiled grid are trapped electrically and will be driven to the coaxial collector wire, giving a pressure-dependent current into the electrometer measurement circuit.

The electrons hit the grid with a significant amount of energy and may cause the generation of soft x-rays. These can impinge on the collector, causing photo-electric emission. This will cause a small amount of current to be sensed even if there are no gas molecules in the gauge. The resultant pressure apparently measured is known as the x-ray pressure limit. The small area of the collector wire limits this effect, so pressures down to 1.3×10^{-6} Pa (10^{-8} torr) can be measured accurately.

Ions collected on the wire give up their charge and re-form as neutral molecules. They may return to the gas phase or stay adsorbed on the collector surface. The ion gauge therefore acts as an ion pump and it is necessary to degas the gauge from time to time. This is usually achieved by using the grid coil as a heater. A large current is passed through the grid coil, which raises the gauge temperature. Heating for 10 to 15 min every week is usually enough to outgas it sufficiently.

Some ions are formed outside the coiled grid; they are repelled and will stick to the gauge enclosure. These external ion/electron pairs usually recombine to form neutral molecules, which may well stick to the gauge's inner surface. As the surface becomes "dirty," significant tracking may occur between the collector grid and filament feedthrough, causing the gauge to give erratic or erroneous readings. This, too, can be overcome by outgassing the gauge by periodically heating the coil for 10 to 15 min.

Vacuum Seals

Mass spectrometer vacuum chambers are not usually made from a single enclosure but rather from a number of parts connected together. To ensure the vacuum integrity of each join, a vacuum seal is required. As with the vacuum chamber, the seals must be impervious to gas flow, and the material of the seal must be soft enough to match exactly to the mating surfaces. It should also show a low vapor pressure and a low rate of outgassing, so that there is no significant contribution to the residual gas in the system.

Seals can be divided into two basic categories, demountable and moving. The material used for each type will depend on the exact function and operating conditions of the seal.

Demountable Seals. Some parts of the vacuum system are joined semipermanently, but it is necessary to open up or take apart sections from time to time for routine servicing and repairs. If this were not the case, then the joined parts could be welded together. "Rubber" O-rings are commonly used, especially for medium- to high-vacuum applications. A number of different elastomers have been tried, and the major problem with them is the outgassing of hydrocarbons. For GC/MS work a compromise between the partial pressure and the nature of the outgassed material is necessary. To minimize the effect on system performance it is necessary for the outgassed material to be of low molecular weight. This becomes even more critical if the seal is to be operated above room temperature.

The most common rubberlike material used for O-rings is Viton. At room temperature, Viton tends to outgas water, carbon monoxide, and carbin dioxide with very little hydrocarbon. As the temperature rises, the hydrocarbon levels also rise, but even so the seal is satisfactory for most applications up to almost 100°C and can be used in systems that have to be baked (outgassed by heating) up to about 300°C. These properties have led to an increasing use of Viton seals in GC/MS vacuum systems in recent years.

For lower backgrounds at elevated temperatures and for the operation at very high or ultrahigh vacuum, it is necessary to use metal seals. These have the disadvantage of having to be replaced each time the seal is opened, although with care some knife-edge flange gaskets can be reused several times. In this type of seal, the surfaces to be joined are machined so as to form sharp projections or knife edges. These cut into either side of the metal gasket which is formed into a flat ring. Oxygen-free, high conductivity (OFHC) copper, a soft, high quality copper, is usually used. When opening up the chamber or replacing an old copper gasket, it is often necessary to pry the pieces apart. Care must be taken not to damage the small knife edges of the flanges.

O-rings and knife-edge flanges do not allow the joined parts to mate precisely, and the actual alignment will depend on the tightness of the flange bolt and the size of the seal. In areas where alignment is critical, spigot flanges are often used. The two joining pieces are carefully machined to give a precise fit. A very thin ring of soft metal is used to effect a seal, gold and indium being the most common. These rings, only a fraction of a millimeter in diameter, are compressed almost to nothing as the flanges are bolted together and the precise alignment is maintained. Indium and gold rings are not reusable, although the material is usually saved as valuable scrap.

Figure 12 compares the various types of demountable seals.

Moving Seals. Vacuum interlock valves and solids probe inlet seals must show properties similar to those of demountable seals, as well as being resilient enough to form a seal while moving (or to a moving part, such as a probe wand). Viton and Teflon are the most commonly used materials. For connections to a very high vacuum, a single seal is not usually sufficient, the leak rate during motion being too great. In order to overcome this problem, two or three seals are used in series, as shown in Figure 13. Often the space between the seals is pumped independently as this greatly reduces the leak rate to the high-vacuum enclosure.

Wear is usually high, and these seals tend to have a short life. Teflon, in particular, tends to flow, and seals made from it have to be replaced regularly. In

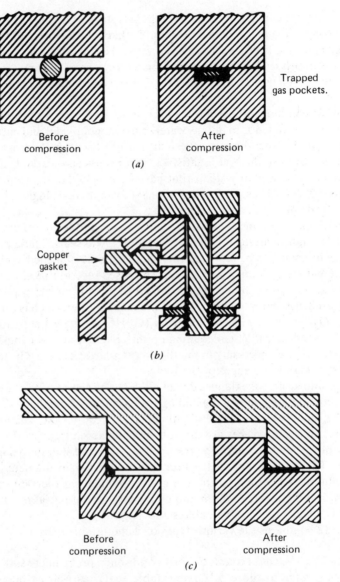

Figure 12. Three types of demountable vacuum seals. (*a*) O-ring in retaining groove. Note how trapped gas pockets are minimized by correct ratio of O-ring to groove size. (*b*) Knife-edge seal. (*c*) Spigot flange. Gold or indium ring (seen in cross section here) compresses to a very thin seal.

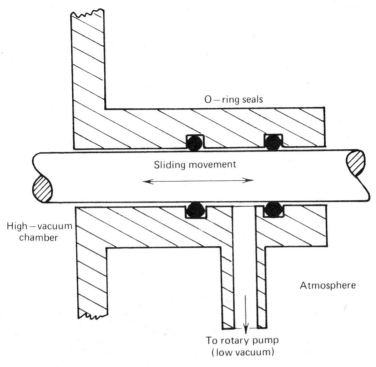

O—ring seals

Sliding movement

High—vacuum
chamber

Atmosphere

To rotary pump
(low vacuum)

Figure 13.   Differentially pumped sliding seal.

some cases, Viton seals can be lubricated. The lubricating fluid must show the same high qualities as the rest of the vacuum chamber. Apiezon greases and even diffusion pump oil can be used. Only the smallest quantities should be applied, or there will be a serious risk of system contamination. Excess grease or oil can also trap gas, giving rise to virtual leaks.

Valve Requirements in GC/MS

Valves may be used to separate different parts of the system. Isolation valves above diffusion pumps, for instance, allow the operator to vent the manifold region to atmosphere without switching off and cooling the pump heaters. This is a great time-saver, as the pumps may take up to an hour to cool and a similar time to warm up. The differential pressure across the isolation valve is often used to good advantage to hold the valve firmly closed; this allows the use of relatively hard and therefore long-lasting seal materials. Before the diffusion pump isolation valves are opened, the manifold must be evacuated. A valve between the manifold and a rotary pump is opened to reduce the pressure sufficiently. This valve must be closed before the isolation valves are opened, or significant rotary pump oil backstreaming will occur.

It is also necessary to allow objects or gases into the vacuum chamber. Vacuum interlocks usually have two valves, one connected to a rough pumping line and another allowing entry to the chamber. In the case of a probe inlet, the probe wand is positioned and an outer seal is tightened. The trapped air is then pumped away by opening the valve to a rotary pump. When the pressure is low enough, the valve to the main vacuum chamber can be opened and the probe inserted. There is usually a significant leak as the probe wand slides through the seal. A second seal at the high-vacuum side of the assembly helps minimize the effect of such a leak.

Gases may also be admitted to the vacuum system from batch inlets or calibration sample vials, or from chemical ionization (CI) reagent gas inlets. The valves used for this purpose must provide both the on/off and flow-regulation functions. Valves that are satisfactory for carrying gases or liquids under pressure may not work well when the pressure differential is reversed. It is important that the valve be chosen specifically for vacuum applications with due consideration to the contribution made to the background spectrum by the valve materials. It is also important to choose a valve that can handle the gases used. Some

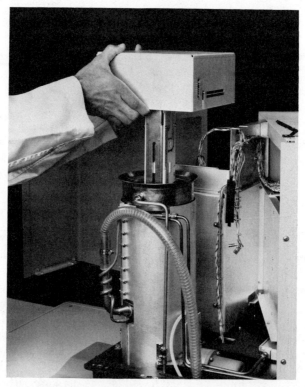

Figure 14. A novel solution to GC/MS vacuum system design. The HP 5992 mass analyzer is shown being placed inside its own diffusion pump. (Hewlett-Packard Corporation)

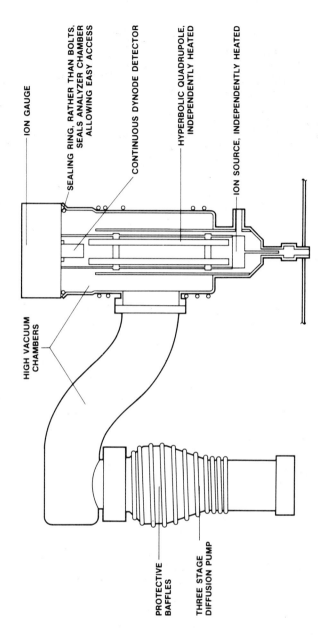

ION GAUGE

SEALING RING, RATHER THAN BOLTS,
SEALS ANALYZER CHAMBER
ALLOWING EASY ACCESS

CONTINUOUS DYNODE DETECTOR

HYPERBOLIC QUADRUPOLE,
INDEPENDENTLY HEATED

ION SOURCE, INDEPENDENTLY HEATED

HIGH VACUUM
CHAMBERS

PROTECTIVE
BAFFLES

THREE STAGE
DIFFUSION PUMP

Figure 15. The compact HP 5995B vacuum system arrangement. (Hewlett-Packard Corporation)

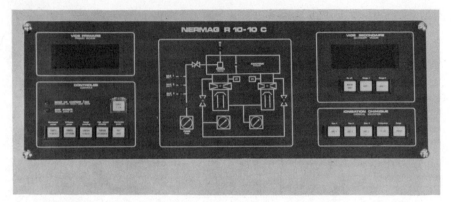

Figure 16.   Vacuum system and CI reagent gas controller. (Nermag SA)

CI reagent gases are very corrosive, and the correct type of valve metal body and synthetic seals must be specified.

Other Pumping Arrangements

The pumping arrangements discussed so far are by no means the only approaches manufacturers have used on their GC/MS systems. Hewlett-Packard, with their bench-top GC/MS, actually placed their mass spectrometer assembly inside a diffusion pump (Fig. 14). This gives a very compact arrangement, truly a bench-top instrument.

A criticism of this design has been that the transfer line from the gas chromatograph must pass through the hot diffusion pump oil at the pump base. This problem is overcome by retaining the basic tight enclosure around the mass spectrometer but using an external diffusion pump (Fig. 15). The package is still quite small and is easily accommodated in a bench-top unit. The transfer line and ion source can be independently heated, an important consideration for use in the chemical ionization mode.

An interesting feature of two instruments described above is the mass analyzer flange seal. An elastomer O-ring sealing against a flared mouth is used. The vertical mounting arrangement makes use of the analyzer's weight as well as the vacuum to hold it in place. As a consequence, bolts are not required to secure the assembly.

Figure 16 shows the vacuum system electronics controller from a Nermag R 10-10 GC/MS system. Most GC/MS instruments have similar control boxes either as a composite unit or as separate valve, gas, and gauge controllers. The unusual feature in this case is the lack of an interface unit between the gas chromatograph and the mass spectrometer. The direct connection allows maximum sample transfer to the mass spectrometer. The ion source must be differentially pumped relative to the analyzer by quite a high-speed diffusion pump to maintain the required vacuum pressures.

CHAPTER THREE

Mass Spectrometers

The purpose of the mass spectrometer part of a GC/MS system is to provide some definite information about the compounds as they elute from the gas chromatograph. In general terms, the GC/MS has to perform one of two tasks: either identification of unknown compounds or the detection of a known compound that may or may not be present. The mass spectrometer provides information necessary to fulfill these functions.

A compound's mass spectrum is a unique chemical fingerprint. It may be possible to determine the compound's molecular weight from its spectrum. The cracking pattern resulting from collision-induced fragmentation in the ion source gives the analyst information about the compound's structure. It may therefore be possible to either identify an unknown compound or at least place it within a chemical class. Once a spectrum is known, certain features of it may be recognized as being particularly representative of the compound. The mass spectrometer can be operated so as to select these features, and if they are detected then the presence of the known compound can be confirmed. Furthermore, integration of the mass spectrum will give an output trace that is indicative of the amount of compound present. With suitable techniques, accurate quantitation is possible even at very low levels.

MASS SPECTROMETER TERMS AND ABBREVIATIONS

α	Angular spread of the ion beam leaving the source of a magnetic sector instrument.
Analyzer	The mass-separating part of a mass spectrometer.
β	Energy spread of the ion beam in a sector instrument.
Calibration compound	A chemical introduced into the GC/MS system for the purposes of tuning, adjustment, alignment, and mass calibration.

CDEM	Continuous dynode electron multiplier.
CI	Chemical ionization.
Collector	On sector instruments, the detector end of the mass spectrometer where the mass-selected ion beam signal is collected.
	On quadrupole instruments the collector is an element of the ion source mounted opposite the filament used to draw the electron beam through the ion source. On sector instruments this plate is usually called the trap.
Daughter ion	Ion formed as a result of fragmentation of another ion, which is referred to as the *parent ion*.
DCI	Desorption chemical ionization. Ionization by a reagent gas of material introduced to the ion source on a specially designed probe.
DFTPP	Decafluorotriphenyl phosphate, a calibration compound used to evaluate GC/MS performance.
Electron energy	Energy gained by an electron as it moves from the ion source filament to the ion volume.
EI	Electron impact ionization.
ESA	Electrostatic analyzer.
eV	Electron volt. Energy gained by an electron or ion of unit charge when it falls through a potential of 1 V.
Extractor	A name given to one or more of the lenses in the ion source.
FC43	Trade name for Perfluorotributylamine, sometimes also known as PFTBA, which is used as a calibration compound.
FD	Field desorption. Ionization of a nonvolatile substance from a specially prepared surface under the influence of an applied electric field.
FI	Field ionization. Ionization of a compound under the influence of an intense electric field.
GC/MS	Gas chromatograph/mass spectrometer, the combined instrument.
Ion energy	Final or total kinetic energy gained by ions as they travel from the ion source to the mass analyzer. Usually expressed in electron volts (or volts for singly charged ions).
Ionizer	Part of the mass spectrometer where ionization and ion focusing are achieved. Also referred to as *ion source* or *source*.
Ion volume	The part of the ion source where ionization takes place.
LBO	Light beam oscillographic recorder, another name for a signal recorder using UV-sensitive paper.

Lens	A plate in the source with a slit or hole through which the ion beam passes. Voltages applied to the lenses control the shape (focus) of the ion beam.
Metastable ion	Ion that decomposes during its passage through the mass spectrometer.
MIM	Multiple ion monitoring.
mmu	Milimass unit.
Molecular adduct ion	An ion formed by the addition of an ionized species to a molecule to yield an ion of higher mass.
m/e, m/z	Mass-to-charge ratio. m/e is valid only for a singly charged ion and is often used incorrectly. m/z for singly and multiply charged ions is the accepted form.
Molecular ion	An ionized molecule that is neither fragmented in the ionization process nor has massive entities added to it.
MS	Mass spectrometer.
Parent ion	An ion from which other ions are derived as a result of fragmentation. The parent ion will be the molecular ion for the primary fragmentation process.
PFK	Perfluoronated kerosene, a calibration compound.
PPINICI	Pulsed positive ion, negative ion, chemical ionization.
Proton affinity	The negative of the enthalpy change for a protonizing reaction; related to the ability of a molecule to capture a proton in certain ionization reactions.
Pseudomolecular ion	A molecular adduct ion. In some cases, such as hydride extraction, an ion close to the molecular weight is formed by a small mass loss. This, too, is sometimes called a pseudomolecular ion if the mechanism is not direct fragmentation.
Pyraprobe	Insertion probe used for mass spectral analysis of pyrolysis products.
Quadrupole (quad)	Type of mass spectrometer.
Quad offset	Ion energy voltage. The term *quad offset* is used on quadrupole instruments where the ion source is operated at ground potential. The ion energy voltage drop is achieved by offsetting all the quadrupole rods by an equal voltage.
RIC	Reconstructed ion current. Total ion current obtained by summing the ion currents of all ions in a spectrum.
Repeller	An electrode mounted inside the ion volume to which a voltage is applied so as to push out the ions formed in the source.
Resolution	The parameter that defines the mass spectrometer's ability to distinguish ions of different mass. Usually defined in

terms of the mass at which adjacent (nominal mass) ions of equal amplitude are separated to give a set level or valley between them, and known as the *percentage valley definition*. Alternatively defined in terms of a mass ion and its width at a given height, and known as the *percentage height definition*. In general terms, resolution = $m/\Delta m$.

Sector instrument	Type of mass spectrometer employing a magnetic field sector and possibly but not necessarily an electrostatic sector.
SIM	Single ion monitoring or selected ion monitoring, depending on the context.
Slit	A gap or hole in a metal plate that defines the beam width or height in a magnetic sector mass spectrometer. Often adjustable in size.
TIC	Total ion current, or total ion chromatogram.
Trap	Electrode that receives the current from the filament after it passes through the ionizer in sector instruments. Called the *collector* in quadrupole instruments.
u	Preferred abbreviation for atomic mass unit.
UV recorder	A recorder that uses mirror galvanometers to focus an ultraviolet (UV) light source onto photosensitive paper to record an input signal.
z	Length (or height) of ion beam measured in plane of magnetic field in magnetic sector instruments.

BASIC MASS SPECTROMETER DESCRIPTION

Before the mass spectrometer can analyze a sample it is necessary that the sample molecule be ionized. The magnetic and electric fields of the mass spectrometer interact with the charged molecules, separating ions of different mass-to-charge ratios. A number of techniques have been used to impart the charge on the molecules. The most common method employed is *electron impact ionization, EI*.

In an EI source, electrons from a heated filament are accelerated across the ionization chamber, which is sometimes called the ion volume. The effluent from the gas chromatograph also passes through the ionization chamber. The electrons interact with these gas molecules, transferring energy to them in the inelastic collisions that take place. If sufficient energy is transferred, the molecule will become significantly excited and may release an electron, giving rise to a molecular ion. In many cases the energy given to the molecule is such that it enters a very excited state and may break up. The resultant fragments may also be electrically charged. The reactions by which these ions are formed are complex and will depend on a number of parameters, not least the nature of the parent molecule itself.

The charged fragments are drawn by the electrical fields in the ion source into the analyzer section, and here they are separated according to their mass-to-charge ratios. Under scan conditions the masses leaving the analyzer are time-dependent, that is, they are selected in turn by the analyzer to be presented to the detector. Each ion normally carries only one electrical charge. The movement of these charges is equivalent to a current flowing. The current levels are usually very small, typically 10^{-11} to 10^{-10} A. Some form of amplification is needed to detect these extremely low levels. This usually takes the form of an electron multiplier giving a typical gain of 10^5. The multiplier is followed by an electronic amplifier or electrometer that gives a current-to-voltage conversion of about 10^7 V/A. The output of the electrometer is then fed to further amplification, filter, and manipulation stages. These form the electronic signal-processing part of the GC/MS instrument and may well end with a computer system.

The main elements of a mass spectrometer are illustrated in Figure 17, the details of which are discussed in the following sections.

IONIZERS

All types of mass spectrometer require that the molecules for analysis first be ionized. In GC/MS instruments the ionizer incorporates lens plates to focus the ion beam as well as to extract and accelerate the ions into the analyzer. The voltage requirements of the lens plates depend on the type of mass spectrometer, but the basic mode of operation for different types of ion source is the same for both sector and quadrupole instruments.

Electron Impact Sources

Electrons produced from a heated filament wire are accelerated through a chamber where they undergo inelastic collisions with the molecules of the gas to be

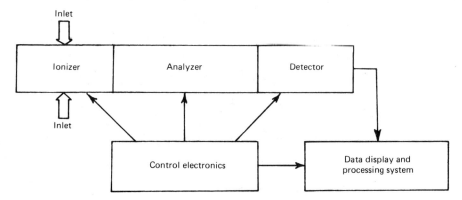

Figure 17. Basic mass spectrometer system.

analyzed. The gas molecules ionized as a result of these collisions are drawn out of the ionization chamber, focused, and injected into the mass analyzer by a series of electrically charged plates. The source may also contain charged plates to mop up excess electrons as well as grounded plates for defining the size of the emerging ion beam. A cross section of a typical electron impact source is shown in Figure 18.

The filament is heated to red or even white heat by passing a relatively large current through it. Electrons are released from the hot surface by thermal emission. The filament material is chosen to maximize the efficiency of this process, rhenium and rhenium alloys being most commonly used. The electrons are accelerated toward the ion volume by applying a voltage known as the electron energy voltage between it and the filament. The energy given to an electron as it passes to the ion volume can be expressed in electron volts, the electron charge multiplied by the electron energy voltage for each electron.

Some but not all of the electrons enter the ionization chamber or ion volume, where they interact with any gas molecules present. The gas may come from the GC/MS interface, probe inlet, or other sample inlet, as well as from the residual gas in the vacuum system. Collisions take place, and energy is transferred from the electrons to the gas molecules. If the electron energy is great enough there is a high probability that some of the gas molecules will become sufficiently excited to be ionized. If the electron energy is further increased, then the molecules may become so excited that they break apart, yielding a number of different types of fragments. The principal reactions can be summarized by considering the effect of fast (high energy) electrons on a hypothetical molecule ABC (Table 1). Some, or even all, of the modes of ionization described in Table 1 may take place in the ion source at the same time for a given compound. However, some forms will predominate, giving rise to a characteristic or *"fingerprint"* spectrum.

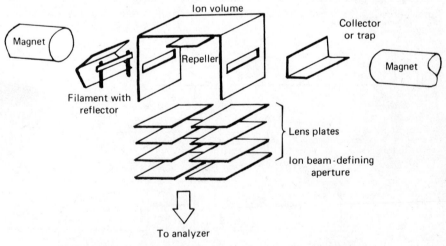

Figure 18. Cross section of an electron impact ion source.

TABLE 1
Electron Impact Ionization Mechanisms. The Effect of a Fast Electron Colliding with the Hypothetical Molecule ABC

Ionization Mechanism	Ionization Products[a,b]			
Molecular ion formation	ABC^+	$+ 2e_s^-$		
Multiply charged ions	ABC^{2+}	$+ 3e_s^-$		
Fragmentation with charged	AB^+	$+ C$	$+ 2e_s^-$	
and neutral fragments	A^+	$+ BC$	$+ 2e_s^-$	
	A^+	$+ B^+$	$+ C^-$	$+ 3e_s^-$
Rearrangement	AC^+	$+ B$	$+ 2e_s^-$	
	AC	$+ B^+$	$+ 2e_s^-$	
Electron-capture, molecular	ABC^-			
ion formation, and	AB^-	$+ C$		
fragmentation	A^-	$+ BC$		
Ion-pair formation	A^+	$+ BC^-$	$+ e_s^-$	
	AB^-	$+ C^+$	$+ e_s^-$	

[a] e_s^- represents a slow, low-energy electron.
[b] Typical ionization reactions are shown; others are possible.

Most organic compounds give more positive than negative ions when ionization occurs under EI conditions. Traditionally GC/MS systems have been configured to examine the positive-ion beam. In this mode it is necessary to remove the slow electrons and excess fast electrons from the ion volume; otherwise recombination may occur. Also, the buildup of negative charges in the ion source may prevent electron emission from the filament, a space-charge effect. A plate or electrode with a small positive potential relative to the ion volume is used to mop up these electrons. In magnetic instruments this is usually called the *trap*; in quadrupoles, the *collector*. It is mounted just outside the ion volume on the side opposite the filament.

The ionization efficiency of the ion source should be as high as possible for maximum system sensitivity. Ionization depends on the chance interaction of an electron with a gas molecule. To increase the probability of this collision we can increase the number of electrons entering the ion source by increasing the electron emission current. This is achieved by increasing filament heating current and raising the filament temperature. The process cannot be continued indefinitely because the cloud of electrons formed above the hot filament surface reduces the emission efficiency due to the space charge formed.

Increasing the electron residence time in the ion volume, which will also increase ionization efficiency, can be done by making the ionization chamber larger but may lead to focusing problems. Electron residence time may also be increased by reducing the velocity. The electrons can be slowed by reducing the electron energy voltage, but, of course, this reduces the energy transfer. The probability of a collision is increased, but insufficient energy is transferred for ionization. However, at some intermediate energy, the ionization potential for the molecule may just be reached in some collisions. Although total ionization and therefore overall sensitivity are reduced, the ionized molecule may have insufficient energy for fragmentation. Consequently, there is a relative enhancement of the molecular or parent ion, an effect that may be used to advantage in gaining vital information about the sample under analysis.

The electron entrance and exit slits in the ion volume are small so that the ions formed are trapped within a charged enclosure (sometimes called a Faraday cage, after Michael Faraday, who first observed that charges formed within an enclosure will assume the same potential as the enclosure). Some electrons are unable to penetrate the ion volume and so hit the outside. If a magnetic field is applied along the direction of the electron beam, electrons traveling along the field into the ion volume will be unaffected. Electrons traveling at an angle, however, will have a component of motion at right angles to the field and will behave as if they were tiny motors. That is, they will experience a force at right angles to both the magnetic field and the direction of travel. This force causes the electrons to move in a spiral and limits the outward spread of the electron beam. It also causes the electrons to sweep out a larger volume and as a result increases the probability of a collision with a gas molecule.

The electron beam is also focused into the ion volume by the use of a filament reflector. A metal reflector plate is positioned behind the filament and is electrically connected to the most negative end of it. Electrons leaving the filament will be repelled from the reflector, which, if suitably shaped, may help to focus the electron beam toward the ion-volume entrance slit.

For positive ion detection the ion volume is normally held at positive voltage with respect to ground. This is often referred to as the *ion energy voltage*. The ions formed inside will adopt the same potential as in the ion volume. Passing from the ion volume toward grounded exit slit or orifice causes the ions to lose potential energy and to gain kinetic energy. A series of charged plates between the ion volume and the exit plate improve the ion extraction and thus increase the source efficiency. The plates also limit the angular spread of the emerging ion beam. In other words, they act as a lens, focusing the ion beam. It is for this reason that the plates are sometimes called *lens plates*, but the plate nearest the ion volume is often called the *extractor*. The number of lens plates varies from one source design to another.

Some sources have a repeller plate mounted inside the ion volume. This normally has a small positive voltage with respect to the ion volume and causes the positive ions to be pushed out or repelled from it. Sometimes tuning with a slightly negative voltage with respect to the ion volume gives better sensitivity. The reason for this will depend on the source design but is usually due to the repeller's effect on the electron beam. In this mode ions are formed farther away from the lens plates. They therefore travel farther as they are being pulled out of the source and may be more sharply focused.

Although the ion volume is normally held at some voltage with respect to earth, the important thing is that the ions should fall through some potential and thus gather kinetic energy. In some quadrupole instruments the rods are all off-set by the same dc voltage so that the mean voltage at their entrance is negative. The ion volume is grounded and the ions are accelerated out and focused in the conventional manner. The potential drop is referred to as the *quad. offset* instead of ion energy. This arrangement allows mechanical connection to the ion source without the need for complex electrical isolation. A typical ion energy voltage for a quadrupole instrument is 8 V. Magnetic instruments operate at much higher voltages; 4000 V is typical.

For negative ion work the ion energy (quad. offset), lens, extractor, and repeller voltages must be reversed in polarity. The analyzers of magnetic sector instruments and the detectors of quadrupole instruments must also be further modified. The collector or trap is not required and is usually connected electrically to the ion volume. Negative-ion work is more common with chemical ionization sources.

Chemical Ionization Sources

If molecules could be ionized without a significant addition of energy, fragmentation would be reduced and molecular ions would be much more abundant. Low-energy E1 spectra show a relative increase in the molecular ion amplitude but an overall sensitivity loss. Chemical ionization (CI) can yield more abundant molecular ions without an overall loss of sensitivity. Basically the sample molecule is surrounded by reactive reagent ions and charge is passed to the molecule by a chemical interaction without much energy transfer. Protons or even larger charged entitites may be added to the molecule to give significant pseudo- or quasi-molecular ions. This effect is very dependent on the chemical nature of both the sample and the reagent plasma. Valuable information about a compound's structure can be obtained using different reagent gases.

Reagent gas is mixed in the ion volume of the CI source with the sample gas stream. The relative partial pressures are usually greater than 10,000 to 1, the reagent gas pressure being about 133 Pa (1 torr). The majority of ions are formed from the reagent gas, which is ionized initially by electron impact. These ions may combine with themselves or neutral reagent gas molecules to form the reactive plasma, or they may be active in their own right. This can be illustrated by considering the reactions that occur when methane is used as a reagent gas.

Methane yields the following ions when bombarded with high-energy electrons:

$$CH_4^+, \quad CH_3^+, \quad CH_2^+, \quad CH^+, \quad C^+, \quad H_2^+, \quad H^+$$

In other words, all possible combinations are formed, although some are more abundant than others. In the relatively high pressures of the CI source, typically 67 to 133 Pa (0.5 to 1.0 torr), a series of secondary reactions occur:

$$CH_4^+ + CH_4 \longrightarrow CH_5^+ + CH_3$$

$$CH_3^+ + CH_4 \longrightarrow C_2H_5^+ + H_2$$

$$CH_2^+ + CH_4 \longrightarrow C_2H_3^+ + H_2 + H$$

$$CH^+ + CH_4 \longrightarrow C_2H_2^+ + H_2 + H$$

Some of the reaction products will further react with uncharged methane:

$$C_2H_3^+ + CH_4 \longrightarrow C_3H_5^+ + H_2$$

$$(C_2H_2^+ + CH_4)_n \longrightarrow \text{polymers}$$

All these reactions and others not listed occur at the same time. The most prominent ions formed, accounting for approximately 95% of the total ionization, are CH_5^+, $C_2H_5^+$, and $C_3H_5^+$. They are extremely reactive and may attack the sample molecules, passing charge to them. Furthermore, a proton or even the whole reagent ion may become attached to the sample molecule giving $(M + 1)^+$, $(M + 29)^+$, and $(M + 41)^+$ ions, where M is the molecular weight of the sample compound. Often the protonated and addition ions have greater stability than molecular ions formed in the EI mode, so that positive identification of molecular weight may be possible with CI even where there is no molecular ion under EI conditions.

The high gas pressure in the ion volume puts a number of constraints on CI source design. All openings of the source must be very small to limit the amount of gas escaping to the vacuum system. This may limit the size of the source slit or exit aperture and affect overall sensitivity. The pumping system must be specially designed to cope with the relatively large gas flows. The gas rushing out of the source can be used to advantage in extracting the sample ions, although reagent ions will also be lost. In quadrupole instruments, where the required ion energy is relatively low, the gas flow may impart sufficient kinetic energy to the ions without the use of an ion energy voltage. The ion beam is focused by a series of lens plates as in the EI source. The lens structure must allow the excess reagent gas to be pumped away, and it must show a relatively high conductance to the source area pumping system, for otherwise a large amount of gas would enter the analyzer region.

The high source pressure also limits the passage of electrons into the ion volume. In chemical ionization, therefore, it is usual to increase the electron energy voltage to overcome the wind of reagent gas blowing out of the source, 100 to 200 V being typical. Inside the ion volume there are so many collisions that the electrons are unlikely to find their way across the source. The motion of all the charged particles within the ion volume is primarily a function of the gas pressure and flow and not the electrical potentials. The collector or trap is not required, and if it is fitted it is usually held at the same potential as the ion volume for CI operation.

Chemical ionization is a "softer" ionization technique than electron impact ionization. That is to say, there is less energy transferred to the sample molecule and consequently less fragmentation. The degree of fragmentation depends on the chemical nature of the sample, the reagent gas, and the source temperature. In some applications it is usual to keep the source as cool as possible during operation. Even the heat from the filament may cause the source to run at too high a temperature, and some form of heat sink is often provided to compensate for this. Running the source at low temperatures does cause condensation of column bleed, sample, and reagent gas products. Since the CI source is exposed to more material than EI sources, the rate of contamination is greater. As a consequence, the operational life of the source before cleaning is required will be shortened. To overcome this problem, a heater may be fitted so that the contaminants can be baked off. Furthermore, the heater can be used in association with a feedback control circuit and the heat sink to hold the source temperature steady. This ensures that spectra for a given compound may be reproduced under consistent conditions.

Reagent Gas Selection.   For ionization to take place at all, the chemical reaction between the sample and the reagent must be exothermic. The greater the heat of the reaction, the more fragmentation that will occur. For a given sample the heat of reaction depends only on the reagent gas used. Consideration of the chemical thermodynamics of ion-molecule reactions leads to the following general statement:

$$\text{Heat of reaction} = \text{PA(reagent gas)} - \text{PA(sample)}$$

where PA represents the proton affinity of the different molecules. The heat of reaction must be negative (the reaction must be exothermic) for ionization to take place.

To develop and prove this statement is beyond the scope of this book. To do so would also be of limited practical value, since proton affinities are known for only a limited range of compounds. If our sample is an unknown, then the proton affinity is also undefined. However, the statement does show that reagent gases of high proton affinity will give lower (less negative) heats of reaction. If ionization occurs (heat of reaction negative), then these reagent gases will give strong molecular or quasi-molecular ions with little fragmentation. Consideration of proton affinities allows the analyst to control fragmentation by selecting different reagent gases. Proton affinities for the more common CI reagent gases are given in Table 2.

When dealing with a complete unknown, it is important to be able to confidently predict the molecular weight from quasi-molecular ions. Although protonation is the principal mode of ionization leading to $M + 1$ ions, hydride extraction yielding $M - 1$ ions is also common. This is especially true with alkanes or molecules with major saturated side chains. If methane is used as the reagent

TABLE 2
Proton Affinities of CI Reagent Gases

Gas	Proton Affinity, kcal/mol
Helium	42
Methane	127
Water vapor	167
Isobutane	195
Ammonia	207

gas, significant addition ions corresponding to $C_2H_5^+$ ($M + 29$) and $C_3H_5^+$ ($M + 41$) additions may be seen. Hence there is usually sufficient information to make a definite identification of the sample's molecular weight.

Methane has a relatively low proton affinity and is more likely to give some ionization with an unknown sample than either isobutane or ammonia. Methane also has the distinct advantage of being a reasonable carrier gas. This means that a sample-enrichment device is not required in the GC/MS interface and direct coupling is possible. The sample losses associated with the interface are consequently eliminated, and sensitivity may be increased.

These advantages make methane the first choice for a general-purpose CI reagent gas, especially when dealing with unknown samples, but there are a few disadvantages that may be significant. The degree of fragmentation is greater than with some other reagent gases, and there is usually a significant CI reagent background spectrum. All CI reagent gases will give a lot of mass peaks at the low end of the spectrum corresponding to the active species formed. Methane also gives a large number of addition ions, quasi-molecular ions, and fragment ions with the system oil background. This interference spectrum may be significant in setting the sensitivity limit of the instrument for some compounds.

Ammonia does not give the same degree of background interference as methane, but its higher proton affinity may prevent ionization. For those compounds that are ionized, the fragmentation is usually less than with methane. Ammonia is a more difficult gas to handle and requires all stainless steel gas pipes, fittings, and valves. It will attack most rubber seals, although Buna-N nitrile has been used successfully. Plastic pipes should not be used. Ammonia is so hygroscopic that it will pull water through even the highest quality tube. Ammonia solution is far more corrosive than dry ammonia gas. Despite these difficulties, it is finding increasing use as a CI reagent gas.

Isobutane has also found widespread use as a general-purpose reagent gas. Like methane it can be used as a carrier gas, but it is less efficient because of its greater molecular weight. Like ammonia it has a higher proton affinity than methane and consequently will give less fragmentation if ionization occurs.

Sometimes it is desirable to obtain an EI spectrum for a sample when the GC/MS system is set up for CI only. Selecting a reagent gas with very low proton affinity will cause a high degree of fragmentation. Some gases, especially the inert gases, have low proton affinities and are often used. The mode of ionization is by charge exchange, and the result is as if a low electron energy EI spectrum were taken. When selecting a reagent gas for this mode, it is more usual to consider its ionization potential rather than proton affinity. The spectrum obtained is approximately equivalent to an EI spectrum taken at an electron energy voltage equal to the ionization potential of the reagent gas. Table 3 gives ionization potentials for a range of charge-exchange-type CI reagent gases.

Selective Ionization. Chemical ionization can also be a powerful technique with known compounds. With the proper choice of reagent gas it is possible to selectively ionize one compound in a complex mixture. This can lead to a significant improvement in the specific detection limit. It can also be used to overcome problems associated with the detection and identification of chemical and stereoisomers. Isomeric differences in molecules account for only very small energy differences. The energy transfer in an EI investigation is usually so great that the information is lost. The softer ionization in CI and the chemical selectivity of the ionization site on the molecule can yield significant information about isomeric differences.

Hence, CI offers a number of advantages over the more conventional EI technique. Sensitivity and specificity can be improved for many compounds. The molecular weight can often be positively identified. There is less fragmentation, yielding simpler, more easily understood spectra. Chemical isomers and stereoisomers can be distinguished. Compounds can be selectively ionized in complex mixtures. The reagent gas may also be used as a carrier gas, allowing direct coupling to the mass spectrometer. The information obtained is additional and complementary to EI spectra. This is illustrated by the spectra of Figures 19 and 20.

All of these advantages do not necessarily hold true at the same time. The analyst must therefore choose the reagent gas to suit the particular purpose. Tables 4 and 5 show a number of gases that have been used as CI reagents, along with their principal reactive ions and some applications. The tables are far from complete but serve to illustrate the range of possibilities. In addition to the gases

TABLE 3
Ionization Potentials of Various Reagent Gases

Reagent Gas	Ionization Potential, eV
Benzene	9.2
Carbon dioxide	13.8
Carbon monoxide	14.0
Nitrogen	15.6
Argon	15.7
Neon	21.6
Helium	24.6

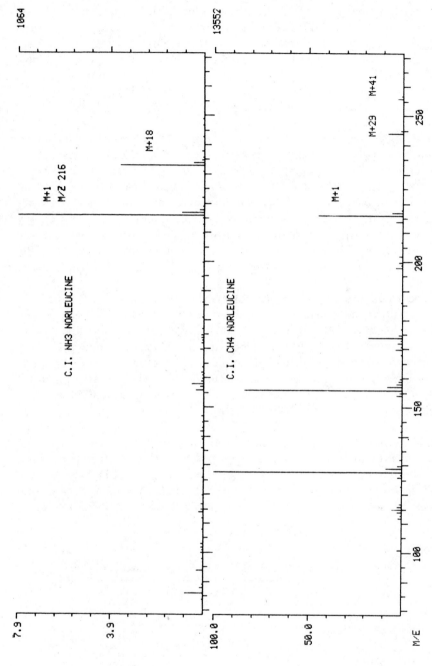

Figure 19. Variations in fragmentation and adduct ions with different CI reagent gases. Norleucine ionized with methane and ammonia (By kind permission of Finnigan-MAT *Ltd.*)

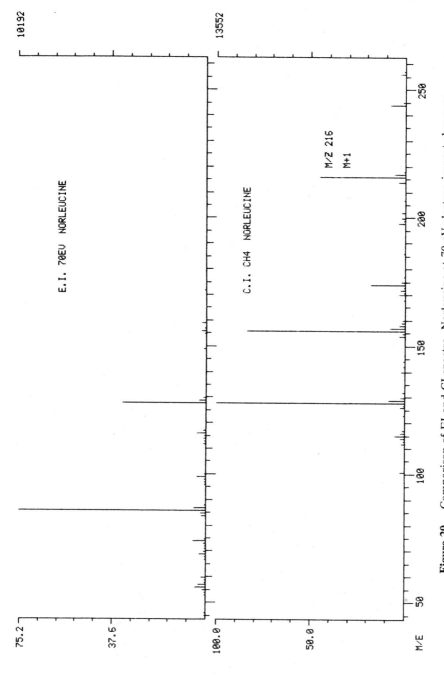

Figure 20. Comparison of EI and CI spectra. Norleucine at 70-eV electron impact shows no molecular ion. Methane CI shows pseudomolecular ions and various ion fragments. (By kind permission of Finnigan-MAT *Ltd.*)

47

TABLE 4
CI Reagent Gas Selection; Positive Ion Mode

Reagent Gas	Reactive Species	Typical Application
Hydrogen	H_3^+	General-purpose positive ion work
Methane	CH_5^+, $C_2H_5^+$, $C_3H_5^+$	
isoButane	$C_4H_9^+$	
Ammonia	NH_4^+, $(NH_3)_2H^+$, $(NH_3)_3H^+$	Polyfunctional compounds; sugars
Water	H_3O^+, $H_5O_2^+$, $H_7O_3^+$	Alcohols, ketones, esters, amines
Nitric oxide	NO^+	Functional group analysis branched hydrocarbons
Deuterated ammonia	ND_4^+, $(ND_3)_2D^+$, $(ND_3)_3D^+$	$1°$, $2°$, $3°$ amines
Deuterated water	D_3O^+, $D_5O_2^+$, $D_7O_3^+$	Active hydrogens
Deuterium	D_3^+	
Helium	He^+	Charge exchange
Nitrogen	N_2^+	
Argon	Ar^+	
Carbon monoxide	CO^+	
Benzene	$C_6H_6^+$	
Neon	Ne^+	

TABLE 5
CI Reagent Gas Selection; Negative-Ion Mode

Reagent Gas	Reactive Species	Typical Application
Methane	Slow electrons	General-purpose negative ion work (simultaneous $+/-$)
Ammonia	Slow electrons	
Hydrogen	H^-	Alcohols
Oxygen	O^-, O_2^-	
Freon 113	Cl^-	Underivatized sugars
Nitrous oxide	OH^-	Simultaneous positive/ negative ion work when mixed with methane
Methyl nitrite	CH_3O^-, HNO^-, NO_2^-	
Carbon dioxides	CO_2^-	Explosives

listed, a number of vapors, for example methanol, have been used. Many of these stem from liquid chromatograph–mass spectrometry (LC/MS) developments. The carrier liquid enters the mass spectrometer ion source, is vaporized, and becomes the reagent for chemical ionization.

Most manufacturers provide several CI gas inlets (Fig. 21) so that two or three reagent gases may be permanently plumbed into the instrument for selection as needed. Provision for pumpout and line purging is usually incorporated, so that fairly rapid changeover from one gas to another is possible. Even alternating CI and EI, ACE, on the same elution gas chromatographic peak is offered by some manufacturers (Figs. 22 and 23).

Negative-Ion CI.   Traditionally, mass spectrometers have been configured to investigate positive ions. For EI work this is reasonable, since most compounds yield a far greater number of positive ions than negative. This statement must be qualified for CI sources; many ion source reactions yield negative ions, and in some cases ions of both polarity are formed. Observing both positive- and negative-ion spectra will give valuable information to the analyst.

To change from the positive to the negative mode requires that the polarity of the ion energy and lens plate voltages be reversed. In quadrupole instruments the analyzer can cope equally well with both polarities, but the ion velocity is too low for the negative ions to penetrate the high negative voltage area at the end of the electron multiplier. The multiplier must be offset by a large positive voltage

Figure 21.   Reagent gas selectors: plumbing arrangments on a CI source. Note also the knob to mechanically switch from EI to CI. (Finnigan-MAT, GmbH)

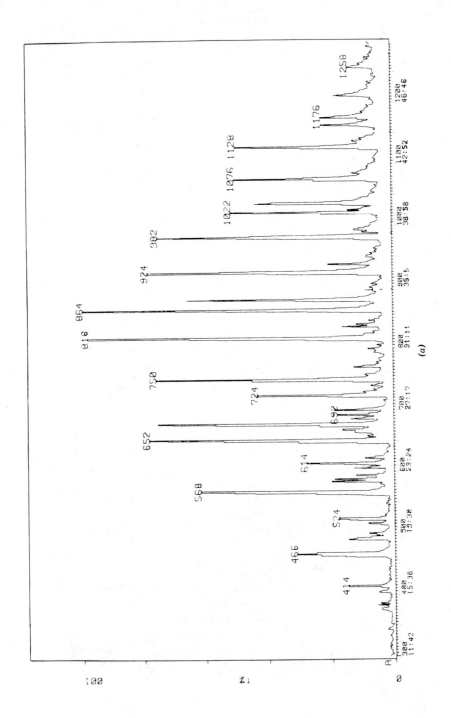

(a)

50

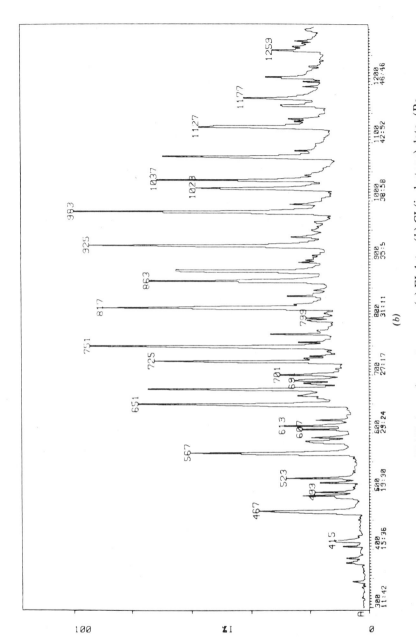

Figure 22. Alternate EI/CI data: chromatograms. (*a*) EI data. (*b*) CI (isobutane) data. (Reproduced by courtesy of VG Analytical Ltd.)

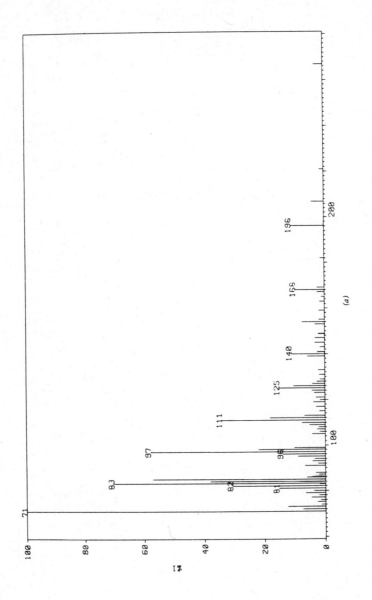

(a)

52

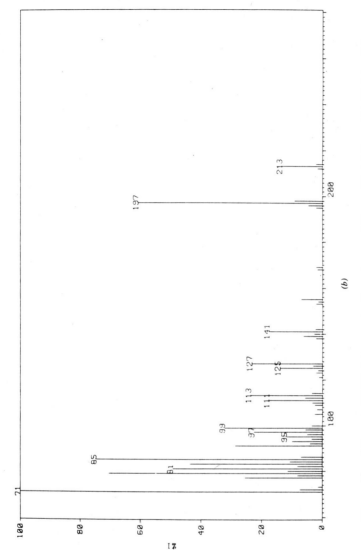

Figure 23. Alternate EI/CI data: spectra from traces of Figure 22. (*a*) Scan 614 EI data. (*b*) Scan 613 CI (isobutane) data. (Reproduced by courtesy of VG Analytical Ltd.)

53

to draw the negative ions in. Translation of the output signal is then a problem, as the input to the signal-processing circuits must be at several kilovolts with respect to ground. This leads to complex circuitry if noise problems are to be avoided. Alternatively, a conversion dynode system must be employed to convert the negative ions into particles with higher energy (usually positive ions) that can overcome the end voltages of the multiplier.

Sector instruments operating with higher ion energies can use a conventional multiplier arrangement, but the analyzer sections must be modified when changing from the positive-ion to the negative-ion mode. The applied voltages in the electrostatic analyzer and the current (and therefore the field) of the magnetic sector must be reversed if the negative ions are to be deflected through the same geometrical arrangement as in the positive-ion mode. As a result, electronic circuits that control the ion flight path are more complex if the mass spectrometer is to be operated in both positive and negative modes.

Since the positive and negative CI data are of significant interest together, it is often desirable to obtain both at the same time or at least on alternate scans of the mass spectrometer. The electromagnet windings of sector instruments are equivalent to very large inductors, and changing current polarity quickly does pose some problems. Recent improvements in magnet design have reduced the inductance, by laminating the core, and switched polarity CI is now possible. Even so, there is a limit to the switching speed, and this will define the minimum sample elution time that can be accommodated.

Quadrupole instruments (and the ESA of double-focusing instruments) are basically electrostatic devices with relatively low stray capacitance and series inductance. With suitable power supply design, high-speed switching is possible either during a scan or on a scan-to-scan basis. The addition of conversion dynodes, one for negative ions and one for positive ions, allows the multiplier to be used for both polarities (Fig. 24). The detector circuitry must be able to separate the data from each mode and process them independently. Even when the data from each mode are to receive the same amplification and filtering, the complexity of the system electronics will be increased by the extra power supplies, switching circuits, and synchronizing arrangement.

Other Ionizer Arrangements

Townsend Discharge Source.   At a relatively high vacuum pressure, about 133 Pa (1 torr), it is possible to strike and sustain a discharge across the gas, creating a highly ionized plasma. A type of source employing this technique, the Townsend discharge source, is illustrated in Figure 30a (see under Dual Sources). It has been found particularly useful in negative-ion CI work with oxidizing reagent gases. A filament is no longer required, and the electron entrance aperture and source can therefore be more tightly closed to prevent sample loss from the ion volume.

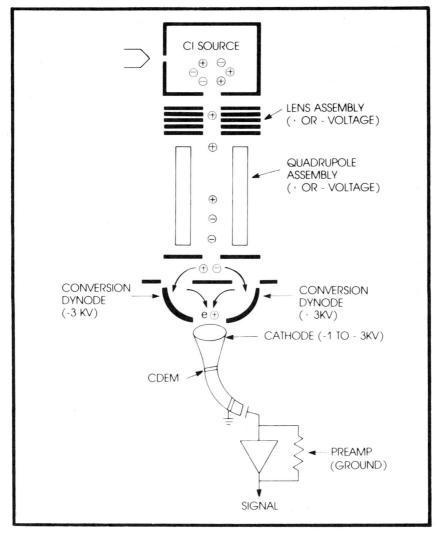

Figure 24. Simultaneous positive-ion and negatives ion detection. (Copyright © 1982 Finnigan Corporation, All rights reserved.)

Field Ionization (FI). This is another mode of ionization sometimes used in GC/MS work. A very high voltage is applied to a sharp point or edge in the ion volume to give a very strong local field, which may be of sufficient strength to pull electrons from the sample molecule. There is little energy transfer to the molecule and thus very little fragmentation. The technique usually yields abundant molecular ions. Unfortunately, the specially prepared emitter surface can absorb contaminants, especially water, which can be ionized and bonded to the sample molecule under the influence of the strong electric field. Consequently, it is sometimes difficult to distinguish the resultant quasi-molecular and addition ions from the molecular ion. Field ionization has not yet been sufficiently developed for widespread acceptance in GC/MS work.

Field Desorption (FD).   This is an ionization mode similar to field ionization, but the sample is actually deposited on the emitter wire (Fig. 25). The technique is not useful in GC/MS work but may be available on the GC/MS system. The emitter assemblies are usually mounted on a probe and can be introduced to the ion source through a vacuum interlock in the same way as a solids probe sample. Figure 26 shows a combined EI/FD source with an emitter assembly on the end of its probe as it is being introduced. Under normal operational conditions this would be done with the system under vacuum. The fact that the emitter can be easily removed, cleaned, or replaced gives FD an advantage compared with FI sources.

Dual Sources.   The analyst will often wish to gather information about a sample using more than one mode of ionization. Electron impact with the large

Figure 25.   Field desorption emitter wire, magnified. (Reproduced by courtesy of VG Analytical Ltd.)

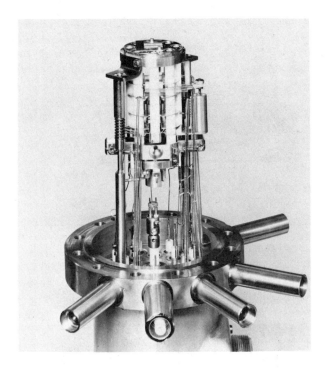

Figure 26. Combined EI/FD source. Note FD emitter on the end of its probe being inserted in the source. (Finnigan-MAT, GmbH)

amount of fragmentation gives valuable structural information. One of the softer ionization methods is often required to determine the sample's molecular weight. It is not always convenient to change the complete source on the system, and a second GC/MS instrument may not be available. Some form of combined source is therefore desirable. EI/CI sources are the most common, but arrangements for EI/FI and EI/FD (Fig. 26) have been produced. Manufacturers have approached the problem of changing over from EI to CI in different ways. The two principal methods used have been to change the source arrangement either mechanically or electrically.

The mechanical changes may allow different pumping speeds, blanking of the collector, and the inclusion or repositioning of the extractor/repeller plate. The Hewlett-Packard 5985B EI/CI source (Fig. 27) illustrates these changes. In the EI mode, a sliding bellows assembly is pulled back. This causes the insulated end piece to make contact with a pin, which not only closes the CI exit aperture but also provides electrical contact for the EI repeller voltage. The electron collector is exposed, and the ionization area is of an open construction. Sliding the selector forward causes the EI repeller assembly to move away from the pin to form the front end of the CI ion volume, with the small exit aperture now open. The pin remains in contact with a second piece of the assembly, which now acts

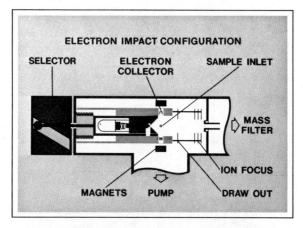

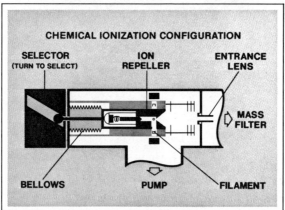

Figure 27. Hewlett-Packard 5985B dual E/CI ion source.

as the ion repeller and forms the rear half of the ion volume. The sliding bellows assembly also blanks off the collector and places a small hole in front of the filament. The ion volume is now a much tighter arrangement.

Varian (now Finnigan MAT) adopted a similar technique for their MAT 44 instrument (Fig. 28). A complete ionization chamber is moved to fit into the EI ionizer when changing over to the CI mode. Each chamber is designed to give optimum performance in the selected ionization mode.

Sources have also been designed to accommodate the change from EI to CI without mechanical variation but by electronic means. The Finnigan 4000 source had five lens plates (Fig. 29). In the EI mode, lenses 1 and 2 give a first-order focus, extracting the ions from the ion chamber. Lenses 3, 4 and 5 refocus the ions into the quadrupole with a much smaller dispersion than lenses 1 and 2. This means that a high proportion of the ions entering the quadrupole satisfy the necessary entrance conditions for successful mass filtering. In the CI mode the

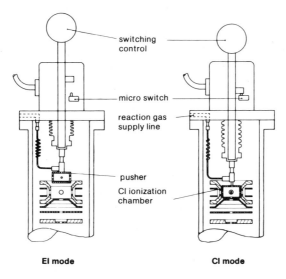

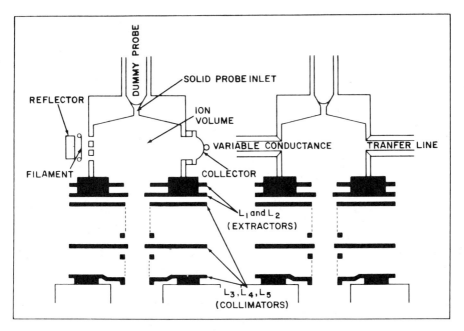

Figure 28. Mechanically switched EI/CI ion source. (Finnigan-MAT, GmbH)

Figure 29. Cross section of The Finnigan MAT 4000 series electronically switchable EI/CI ion source. (Copyright © 1982 Finnigan Corporation. All rights reserved.)

collector and lenses 1 and 2 are set to the same potential as the ion volume and become, in effect, extensions of it. The first two lenses and the solid insulating rings act as a pipe guiding the emerging gas stream toward the quadrupole. The last three lenses are far more open in construction and allow the excess reagent gas to be pumped away. The extensions of lenses 3 and 5 along the source-quadrupole axis shape the field pattern in the source so that the ions are trapped within an apparently continuous charged area. Only a small number of ions escape along with the neutral reagent gas. The ions are focused by the last three lenses and injected into the quadrupole.

Careful design of the electron entrance apertures and the filament mounting overcomes the sensitivity loss normally associated with tight CI sources operated in an EI mode. When operated through a sample-enrichment device, the source pressure remains low enough for EI work. Capillary columns connected directly to the source cause a pressure rise that is too high for normal EI, and charge exchange may take place. To compensate, a plug is pulled back from the source to increase its conductance. The plug is positioned so that the sample gas stream passes across the electron beam before being pumped away. This maintains EI sensitivity and prevents unwanted pressure effects.

It is interesting to note that although the 4000 series source design gave fine results it was prone to contamination under some operating conditions. To overcome this problem and to enhance its features, Finnigan developed the 4500 source (Fig. 30). This has interchangeable ion volumes, which are inserted with a special probe via the solids probe interlock. The different ion volumes are optimized in shape for either EI or CI. The design also allows configuration as a discharge source. This is particularly suitable for CI work with oxidizing reagents that would otherwise burn up filaments.

MASS ANALYZERS

Once ionized, the sample molecules and their fragments can be separated on the basis of their different mass-to-charge ratios. Doubly charged ions will behave in the analyzer as if they were singly charged ions of half the mass. In the discussions that follow, all ions are assumed to be singly charged unless otherwise stated.

The most common forms of mass analyzer used in GC/MS instruments are magnetic sector and quadrupole. Other types of mass spectrometers such as time-of-flight instruments exist, and some have been used in GC/MS applications. They are not generally available commercially as GC/MS instruments, and so discussions have been limited to the two main types.

Magnetic Sector Mass Spectrometer

A conductor placed in a magnetic field will experience a force when current is passed through it. Investigation of this and other electrical phenomena led J. J.

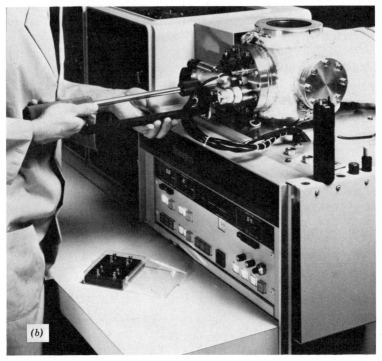

Figure 30. Finnigan-MAT 4500 series EI/CI discharge source. (*a*) Source detail, replacement ion volume about to be inserted in lens block assembly. (*b*) Operators viewpoint, inserting replacement ion volume through the solids probe inlet. (Copyright © 1982, Finnigan Corporation. All rights reserved.)

Thompson and his coworkers to pass an electron beam through low-pressure enclosures subjected to magnetic fields. The initial investigation concerned electrons, but early vacuum techniques were somewhat limited and ionization of the residual gas occurred. It was noted by Thompson and his colleagues that these charged particles could also be deflected by the magnetic field. Furthermore, their relative masses could be assessed by considering the trajectories. Hence, the first magnetic sector mass spectrometer was born.

Instrumentation has come a long way from those early days, but the principles of operation still hold good. If a current flows through a conductor in a magnetic field, then a force will be experienced at right angles to both the field and the direction of the current. The force is defined as

$$F = \mathsf{B}il \tag{1}$$

where B is the field strength, i is the current, and l the length of the conductor.

The direction of the force is given by *Fleming's left-hand rule*. The first three fingers of the left hand are arranged at right angles. The first finger represents the field. The second, center finger represents the current, and the thumb, the resultant thrust (force). Electric current can be defined as the passage of charge. In this case the conductors are the ionized particles themselves. There are no physical restraints on the ions, so they will respond to the exerted force, which will always be at right angles to the direction of travel. Under these conditions the resultant trajectory will be circular. Considering a singly charged ion of mass m, we have

$$i = e/t \ \mathrm{C/s} \tag{2}$$

Hence,

$$F = \mathsf{B}ev \tag{3}$$

where v is the velocity of the ions.

For the ions to travel on a circular path,

$$F = \frac{mv^2}{r} \tag{4}$$

where r is the radius of the trajectory. Hence,

$$\frac{mv^2}{r} = \mathsf{B}ev \tag{5}$$

Armed with this information we can construct a simple mass spectrometer (Fig. 31). Ions are accelerated from the ion source to the magnetic sector by the accelerating voltage V. The spread of the ion beam is limited by the source and

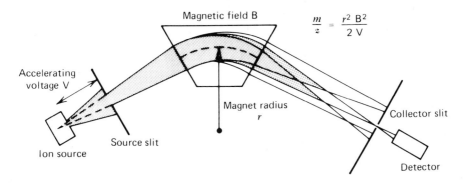

Figure 31. Magnetic sector mass spectrometer.

collector slits and enters and leaves the magnetic field at right angles (beam centerline). The source slit, magnet center of radius, and collector slit are collinear. The velocity of the ion can be calculated from the energy equation of the source/ accelerating voltage, the potential energy lost being equal to the kinetic energy gained so that

$$\tfrac{1}{2}mv^2 = Ve \tag{6}$$

for a singly charged ion. Substituting this into Eq. (5) to eliminate the velocity parameter gives the mass-to-charge ratio as

$$m/e = \frac{r^2B^2}{2V} \tag{7}$$

or in general terms, for an ion of charge z,

$$m/z = \frac{r^2B^2}{2V} \tag{8}$$

Since the electronic charge has been accurately determined in other experiments, the actual mass of the ions in grams could be calculated in terms of the other instrument parameters. However, in practice the important thing is the relative mass selection of this geometrical arrangement. The instrument cannot distinguish between M^+, a singly charged ion, and $(2M)^{2+}$, a doubly charged ion. Masses are normally given relative to each other rather than in absolute units. By international agreement, the most abundant isotope of carbon has been assigned a mass of exactly 12 u.* Masses in a spectrum are quoted on a scale of atomic mass units rather than in absolute units.

*u = atomic mass unit; sometimes abbreviated amu.

Resolution of a Magnetic Sector Mass Spectrometer.   For a mass spectrometer
to be useful, it must be able to distinguish ions of similar but slightly different
mass. To resolve these adjacent masses it is necessary for the collector slit width
to be small so that only a narrow range of masses will be presented to the collector at any one time. Furthermore, the magnet system is analogous to a lens in
optical work, the source slit presenting an image on the collector slit area. It is
not sufficient to have only a narrow collector slit; images of several masses could
overlap the collector slit at the same time. The source slit must also be narrow for
higher resolutions. Reducing the slit sizes will also reduce the magnitude of the
ion beam; that is, sensitivity will decrease with increasing resolution.

Resolution is defined in general terms as the ability of the mass spectrometer
to separate adjacent masses by a given amount. In the 10% valley definition it is
assumed that two masses of exactly equal magnitude 1 u apart are just separated, with an overlap or valley between them of a height equal to 10% of their
amplitude. The resolution figure is then equal to the mass of the first peak. This
is an impossible way to measure resolution except in very rare cases, but the 10%
between these hypothetical mass peaks can be considered the sum of the 5%
heights of each. This leads to a more practical method of measurement.

$$R = \frac{m}{\Delta m} \tag{9}$$

where m is the mass at which resolution is to be measured, and Δm is the mass
peak width at a height half the previously defined valley height, that is, the width
at 5% height for the 10% valley definition. Figure 32 illustrates the relation
between these two methods of definition. The percentage width definition allows
the measurement of resolution at any convenient mass in the spectrum. Since the
resolution of an ideal magnetic mass spectrometer is constant, the resolution is
known for the full mass range and will define the maximum operating mass for a
given percent valley separation.

In the ideal magnetic spectrometer, resolution will depend only on the geometrical arrangement: the slit sizes, magnet radius, and sector angle. However,
mass stability will play an important part in the instrument's ability to separate
adjacent masses. So far we have assumed that the ions all come from a single
point in the ion source with a fixed energy. In practice there will be a range of
ions leaving the source. This range is minimized by the source design, which
includes not only an ionization system but also a series of alignment and focusing
lens plates. Any instability of the instrument parameters, the accelerating voltage V and field strength B will seriously affect the ion beam and, of course, the
resultant resolution. The variation in mass due to the instability of the instrument parameter can be assessed by considering the partial differentiation of
Eq. 8, leading to

$$\frac{\Delta M}{M} = \frac{2\Delta B}{B} - \frac{\Delta V}{V} \tag{10}$$

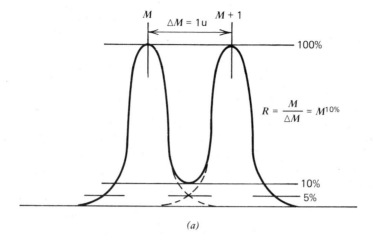

$$R = \frac{M}{\Delta M} = M^{10\%}$$

(a)

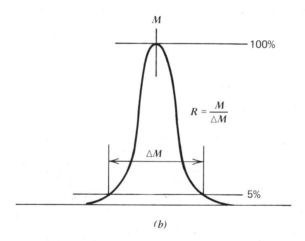

$$R = \frac{M}{\Delta M}$$

(b)

Figure 32.   Equivalence of (a) 10% valley and (b) 5% height resolution definitions.

The field term is doubled because of the squared term in the mass spectrometer equation and is obviously critical to instrument performance. It is for this reason that mass spectrometer magnets tend to be very large units with heavy iron cores and electrical windings of many turns. It is also important that the magnet be stable with temperature. Water cooling and temperature regulation are often employed. Permanent magnets are also possible but are difficult to make in the sizes necessary to accommodate reasonable mass spectrometer flight tubes. The mass selection would also depend entirely on the accelerating voltage. This voltage plays a key role in extracting and focusing the ions from the source and cannot be varied over a wide range without significantly affecting sensitivity.

Examination of Eq. 5 shows that the mass spectrometer equation is one of momentum selection.

$$mv = Ber \qquad (11)$$

Variations in the ion energy will cause ions of the same mass to have different velocities and ions of different mass to have the same momentum. The latter will be selected together and will not be resolved. The energy spread is due in part to instabilities in the accelerating voltage discussed above and due to the initial kinetic energy of the molecules in the source. The variation in the sample's kinetic energy is usually small compared with the energy gained as the ions are accelerated out of the ion source and becomes significant only at higher resolutions.

Scanning Modes for a Magnetic Sector Instrument.   We have seen from Eq. 8 that the mass selected by a magnetic sector mass spectrometer is dependent on the two principal instrument parameters, accelerating voltage and magnetic field. Changing one or both of these parameters will cause the mass spectrometer to scan across a mass range. Cycling the parameters through a range of values will cause a repetitive mass scan. It is obvious from Eq. 8 that the mass selected is a nonlinear function of the instrument parameters. The instrument scanning performance is critically dependent on the manner in which the instrument parameters are varied.

The magnetic field cannot be altered directly; normally the magnetic current is varied. For small values the field is directly proportional to the magnetic current, but as the field increases there is a saturation effect. Further increases in magnetic current give proportionally smaller increases in the magnetic field. Also there are residual magnetism or hysteresis effects. Hence, the field is not always the same for a given magnetic current. Selection of the right scanning mode can minimize problems associated with these effects. Figure 33 compares the scanning modes, which are described below.

Quadratic Scan.   In practical terms it is easier to vary the current in a linear fashion, and therefore the field, providing we avoid high saturating values. Typically,

$$B = B_0(1 + at) \qquad (12)$$

where B_0 is the initial field strength. This leads to a quadratic mass scan (assuming the accelerating voltage is held constant) of the form

$$M = M_0(1 + 2at + a^2t^2) \qquad (13)$$

where M_0 is the initial mass. Although practical, such a scan mode has limitations. Just before the end of the scan, the field is changing at its fastest rate and then suddenly stops. The magnet is basically a large inductor and will resist such

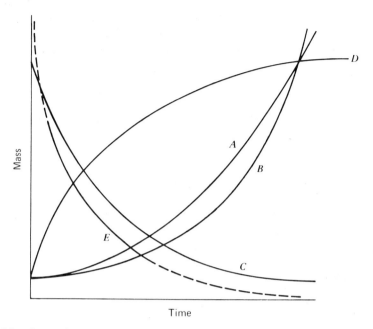

Figure 33. Sector instrument scanning modes. A, Quadratic (linear current) scan; B, exponential scan, up; C, exponential scan, down; D, complementary exponential scan up; E, inverse (linear voltage) scan.

a dramatic change. There will be a significant backlash voltage generated at the end of the scan, which will put the driving circuits under some stress. Moreover, the magnetic field at the end of the scan is liable to wide fluctuation. This situation also occurs with step changes in magnetic current. For instance, as the magnet is reset to the beginning of the scan, the hysteresis effects are quite variable and are a significant problem.

Exponential Scans.   Magnetic scans of the form

$$B = B_0 \exp(bt) \qquad (14)$$

lead to exponential up (b positive) and exponential down (b negative) scans of mass.

$$M = M_0 \exp(2bt) \qquad (15)$$

This type of scan gives a constant time per mass peak (assuming resolution is constant). This is a great advantage from a data-acquisition point of view and is the scanning mode usually used when the data are recorded by a computer system. The exponential up scan has the same problems of changing scan rate as the quadratic scan. Computerized data acquisition is usually made on an exponential down scan following a significantly high mass settling time. Further improvements in stability are possible if the up scan rounds off slowly to the final mass. This is achieved with a complementary exponential up scan of the form

$$M = M_f[1 - a \exp(-ct)] \qquad (16)$$

where M_f is the final high mass.

The start mass M_0 is defined as

$$M_0 = M_f(1 - a) \qquad (17)$$

Complementary exponential up scans are not easy to generate with the normal instrument scanning hardware and are usually attempted only when computer control is available.

Voltage Scans. As previously stated, voltage scanning is usually limited in range to prevent detuning of the ion source. Over a small range it is a useful high-speed scan option and is often used in selected ion monitoring and peak matching. Linear and exponential scans similar in form to those described above can be used. On some computer-controlled instruments, both the accelerating voltage and the magnetic current can be controlled. First, the magnet is parked at some suitable center mass for the nominal accelerating voltage. This is then scanned rapidly about its nominal value to generate a scanned mass window. With this technique the magnet must be precisely set, but setting a magnetic sector instrument to an exact value of mass is difficult because of hysteresis effects.

Magnet Field Control. A given magnet current may give quite a range of masses under different conditions. To overcome this problem, field control of the magnet is often employed. Although the current is altered to change the field (and therefore the mass), the controlling circuits contain a field-sensing device. Various flux meters and Hall-effect devices have been used. They all increase the precision of mass setting but give a reduction in the overall system mass stability.

This apparent contradiction can be explained by considering the typical magnet control block diagram in Figure 34. Under current control the current to the magnet is sensed in a series resistor, and a buffer amplifier returns a proportioned signal. This feedback signal is compared with the input demand signal by an error amplifier. The resultant error signal is used to control the magnet power supply so that current regulation is achieved. With modern electronics this method of negative feedback can have a very rapid response and the magnetic current can be precisely controlled. The resultant field will, of course, depend on the magnet's history and condition and will not be so precise.

Any noise in the system within the feedback loop is rapidly compensated. On the other hand, noise or instability of the control input signal will be faithfully reproduced, giving fluctuations in the magnet power supply. This does not nec-

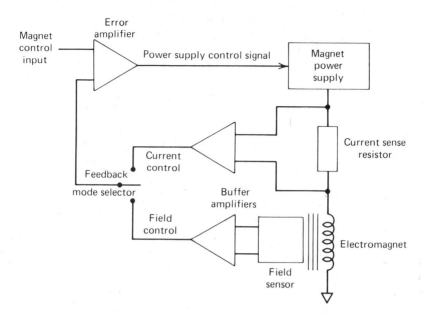

Figure 34.   Electromagnet power supply control.

essarily give rise to significant magnetic field variations. The magnet is equivalent to a very large electrical inductance and will resist all rapid changes. Instabilities and noise that are fast compared with the normal scan speeds are
damped out.

Under field control the feedback signal is taken from a field-sensing device.
The circuits will then try to reproduce the noise and instability of the input signal
at the magnetic field-sensing point. There is no reduction due to the filtering
effects of the magnet's inductance; the control circuits will just drive the power
supplies harder to overcome them. Noise within the control circuits is also a
problem. At frequency levels above the normal scan speeds of the system, there
may be a considerable phase shift from the power supply through the magnet to
the field sensor. To put it another way, there may be a delay in sensing an error
in the field due to noise on the magnet's power supply. By the time a controlling
signal is applied, the system disturbance may have passed. The control signal
will then create a new but opposite disturbance. This is a positive feedback situation and is a very unstable condition. It can be minimized by careful design, but
little can be done to prevent the magnetic field from following any noise on the
input signal when it is under field control.

The magnet's inductance is the most significant parameter in deciding the
maximum scan rate of the magnetic field. Large magnets of large inductance are
desirable for stability and uniformity of the field. To scan these quickly requires
that the power supply be very powerful with a large voltage as well as current
range. Great improvements have been made in recent years by laminating the

iron core of the magnet. This reduces both the inductance and the eddy current heating effect, giving greater speed and thermal stability. Figure 35 shows just such a magnet with a Hall probe field-sensing device fitted between the pole pieces alongside the instrument's flight tube.

Double-Focusing Mass Spectrometers

Examination of the equations for a magnetic sector mass spectrometer shows that it is, in effect, a momentum analyzer. With a relatively high accelerating voltage, the ion energy, and hence the ion velocity, is almost but not quite defined. The velocity will vary slightly due to kinetic energy variations at the time of ionization. These are usually small compared with the effect of the accelerating voltage unless of course, the voltage is reduced appreciably to achieve a high mass range or the ionized material is very hot. Nevertheless, as resolution is increased, these energy variations become significant and limit the maximum resolving power achievable. Furthermore, it is quite difficult to stabilize very high voltages precisely, and some power supply ripple is inevitable. The ripple appears as an energy variation in the ion beam, which will also limit resolution.

To overcome this problem, additional energy-focusing arrangements are used. The most common form is the electrostatic analyzer (ESA), shown in Figure 36, which consists of two curved plates with a voltage between them that is

Figure 35. MM 70-70 laminated magnet assembly. Note field sensor between pole pieces. (Reproduced by courtesy of VG Analytical Ltd.)

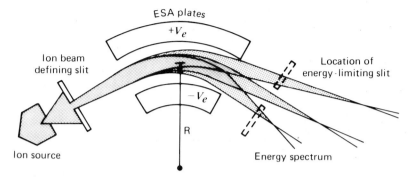

Figure 36. Electrostatic energy analyzer.

high, but not nearly as high as the accelerating voltage; 500 V is typical. The plates are usually driven with one positive and the other negative with respect to ground so as to give zero mean potential along their centerline. Ions entering the electrostatic analyzer will pass through only if

$$\frac{mv^2}{R} = \frac{Ez}{d} \tag{18}$$

where m is the ion mass, v is the velocity, z is the charge on the ion, R is the ESA mean radius, E is the differential ESA voltage, and d is the ESA gap.

This leads to

$$R = \frac{2d}{Ez}(\tfrac{1}{2}mv^2) \tag{19}$$

$$R = \frac{2d}{Ez}U \tag{20}$$

In other words, the electrostatic analyzer is an energy analyzer. Placing a grounded slit before the ESA limits the beam spread and defines the potential accelerating the ions out of the source. A slit after the ESA will limit the range of energies allowed through. The beam passing through this slit can then be the input to a magnetic sector mass analyzer, which will not then be subject to wide energy variations. The equation of motion of an ion passing through the magnetic sector, Eq. 5, can be restated with the substitution

$$U = \tfrac{1}{2}mv^2 \tag{21}$$

representing the kinetic energy of the ion and giving

$$m = \frac{B^2z^2r^2}{2U} \tag{22}$$

Substituting for U from the ESA equation (Eq. 20) leads to

$$m/z = \frac{B^2 r^2 d}{ER} \tag{23}$$

which is independent of the accelerating voltage.

Resolution can be represented by consideration of small changes in the parameters of Eq. 23 so that

$$\frac{1}{R} = \frac{\Delta m}{m}$$

$$= \frac{2\Delta B}{B} + \frac{2\Delta r}{r} + \frac{\Delta d}{d} - \frac{\Delta R}{R} - \frac{\Delta E}{E}$$

The magnet term $\Delta B/B$ appears doubled and is extremely critical. Hence magnet design is aimed at achieving very stable fields. The ESA term $\Delta E/E$ arises from power-supply variations and noise pickup. ESA voltages typically in the range 200 to 800 V are easier to stabilize than the high accelerating voltages of 4 kV and above. The term therefore implies that an improvement in resolution is possible compared with single-sector instruments.

The term $2\Delta r/r - \Delta R/R$ is the implied focusing term and will be zero if the geometry is right. The term $\Delta d/d$ arises from manufacturing tolerances on the ESA plates and from temperature effects and can usually be made very small.

In practice, one does not operate with point sources and detectors. The finite size of source and collector slits tends to limit the available resolution. If the slit sizes are reduced, sensitivity suffers. Also in the arrangement described above, ions of a given mass at a high or low energy are rejected by the slit between the ESA and the magnet, and sensitivity suffers further. However, all is not lost. If the slit between the ESA and the magnet is opened up, more of the ion signal can get through. Ions of higher and lower energies compared to the nominal value enter the magnetic field at a slight angle, while the nominal energy ion enters the magnetic field normally. As a consequence, there is a point where the loci of the focus for mass ions of different velocities and the focus for mass ions entering at different angles cross. This point of double focus is critically dependent on the geometry of the mass spectrometer design. Placing the collector slit at the point of double focus will give the maximum signal for a given mass and will allow higher resolutions than a single-focusing instrument. Figure 37 illustrates the typical double-focusing arrangement. Some designs put the magnet first, which allows the investigation of energy spectra resulting from ion decomposition and ion-gas collisions as well as conventional mass spectrometric and GC/MS work.

Resolution figures vary with design. The Micromass 70-70, for instance, has a resolution limit of about 25,000, which is about 10 times that achievable with a spectrometer based only on the magnet assembly. Resolution figures in excess of 100,000 are possible, but such instruments do not find a lot of use in GC/MS

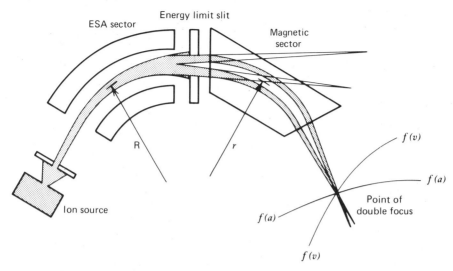

Figure 37. Double-focusing mass spectrometer showing locus of velocity focus $f(v)$ and angular focus $f(a)$. Trajectories are illustrated for ions of a fixed mass but two values of energy.

work because their scan speeds are too slow. Much work has been done to improve scan speeds of instruments with large magnets so that higher resolutions can be applied to GC/MS work. The laminated magnet assembly from a Micromass 70-70 shown in Figure 35 more than doubles the scan speed compared to a standard magnet.

Scanning of double-focusing instruments is the same as for single-focusing instruments; both magnet and voltage scans are possible. In the case of voltage scans, the ESA voltage is scanned. It is normal to derive a reference voltage from the ESA voltage supply to drive the high accelerating voltage supply. Hence, the accelerating voltage scans in sympathy with the ESA voltage, and the limitations described for single-focusing instruments apply here also. With a double-focusing instrument, there is also the possibility of scanning both the electric and magnetic voltages at the same time. These linked scans produce some interesting spectra, especially where ion decompositions occur, and they are useful in the study of metastables and collision-induced fragmentation. Linked scans require careful setting if the performance of the instrument is not to be degraded and are in ideal area of application for data system control. A block diagram of a digital scanner system is shown in Figure 38.

So What Use is High Resolution? At one time it was believed that high resolution was essential to achieve accurate mass assignments. Improvements and the general use of computer control and data-acquisition techniques now mean that accurate assignments can be made on low-resolution spectra providing that the peak shape is good and that there are no interfering mass peaks. Only high reso-

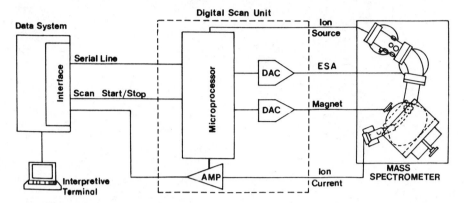

Figure 38. Block diagram of a digital scanner system. (Reproduced by courtesy of VG Analytical Ltd.)

lution can guarantee the latter. Peak matching, comparing the spectrum of an unknown mass to that of a reference mass to measure the mass ration, can give a very accurate measurement of the unknown's mass only if the two peaks are separated and each is separated from other interferences. However, accurate mass measurement using peak matching is a relatively slow process and not easily achieved on spectra from an eluting GC peak.

High resolution is also useful for separating doublets (ions of the same nominal mass). Unfortunately, if the doublet occurs at a relatively high mass, then quite large resolution figures must be used. For instance, to resolve a doublet arising from a difference of ^{13}CH to N in the formulas of two ions requires a resolving power of about 6000 at mass 50 u, 12,000 at mass 100 u, and over 60,000 at 500 u. High resolution, therefore, has limited use in GC/MS work. Ofter higher quality GC performance using capillary columns, high mass spectrometer scan speeds, and sophisticated computerized library routines, all standard features on a modern low-resolution GC/MS system, are preferred.

Nevertheless, some laboratories favor a high-resolution instrument operating at a faster scan speed in a low-resolution mode with the ability to close down the slits to obtain a high-resolution spectrum should the need occur. The usual choice is an instrument that has a resolution limit in the 20,000 to 30,000 region than can also be operated down at the 1000 resolving power level. The Micromass 70-70 system, Figure 39 is a typical example, and the spectrum of Figure 40 shows what can be achieved on a good day, although probably not under GC/MS conditions.

Metastable Response in Sector Instruments

Some of the ions formed in the mass spectrometer ionizer are not stable and will decompose. They may be stable long enough to be drawn out of the ion source and into the mass analyzer. Quadrupole instruments, to be discussed, are con-

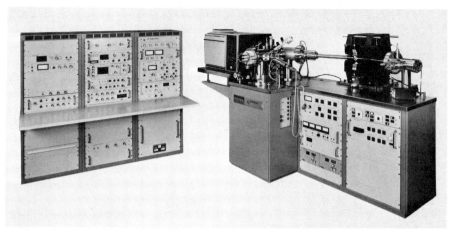

Figure 39. Micromass 70-70 double-focusing mass spectrometer. (Reproduced by courtesy of VG Analytical Ltd.)

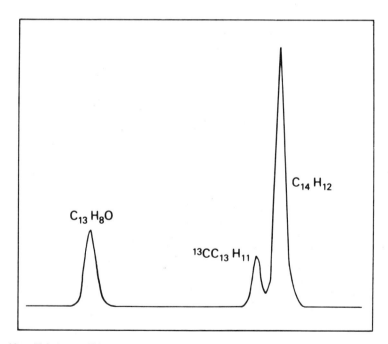

Figure 40. Triplet at 180 u in dibenzosuberone showing partially resolved doublet due to $^{13}CC_{13}H_{11}$ (180.0894) and $C_{14}H_{12}$ (180.0939) plus the completely resolved peak at 180.0575 due to $C_{13}H_8O$. Resolution of doublet shows a resolving power of ~33,000. (Reproduced by courtesy of VG Analytical Ltd.)

tinuous focusing devices and will not allow transmission of these metastables. On the other hand, sector instruments can show a significant response to the ion fragments resulting from a metastable decomposition.

The relatively high accelerating voltages on sector instruments cause the ions to move rapidly from the ion source to the magnetic sector. Decompositions that occur in this region give rise to broad, diffuse peaks in the mass spectrum display as well as the normal ion response. Decompositions at other locations generally lead to trajectories that do not reach the collector. Metastable ions occurring at the magnet entrance give rise to an apparent mass peak m^*,

$$m^* \approx \frac{m_2^2}{m_1} \tag{24}$$

where m_1 is the mass of the original ion and m_2 is the mass of the ion remaining after decomposition.

The exact position of the metastable peak is difficult to determine. The peak is broader than the normal spectral peaks and does not always drop to the baseline. This is illustrated by Figure 41, which shows part of a uv recording of a mass scan. Two metastable peaks are visible, one at about 89.7 u and the other at about 91.3 u.

Although the metastables can give important information to the trained chemist, they are difficult to determine accurately with modern data-acquisition

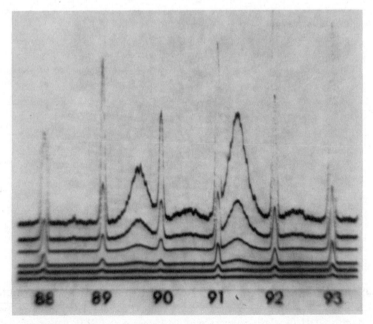

Figure 41. Part of an ultraviolet recording of a spectrum showing metastable ions.

systems. For instance, the data of Figure 41 are clear on the uv traces, but a data system would have difficulty detecting the metastables at low gain. With high gains the metastable peaks would interfere with the normal spectral peaks, and the data system may well merge their signals together. As a result, the use of metastables in sample analysis is less common than it once was.

To overcome the problems that metastables present to data systems, some instruments are fitted with metastable suppressors. Figure 42 shows typical before-and-after traces for part of a mass scan. The result is data that are satisfactory for the computer interface and the peak detection software, but signifi-

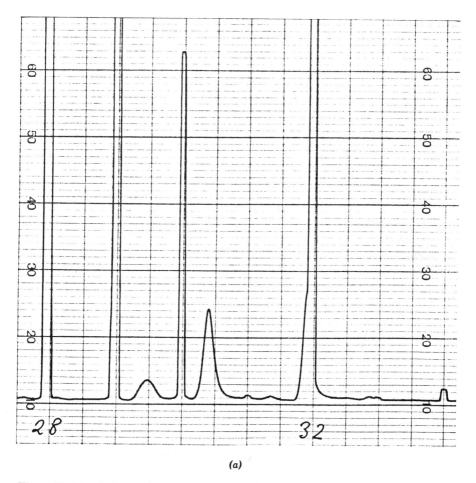

(a)

Figure 42. (a) Stripchart recording of high-resolution spectra showing effect of metastables. The mass range 28 to 34 u is shown for an air–n-butane mixture. Two wider metastable peaks are separately resolved, and a third interferes with the oxygen mass peak at 32 u.

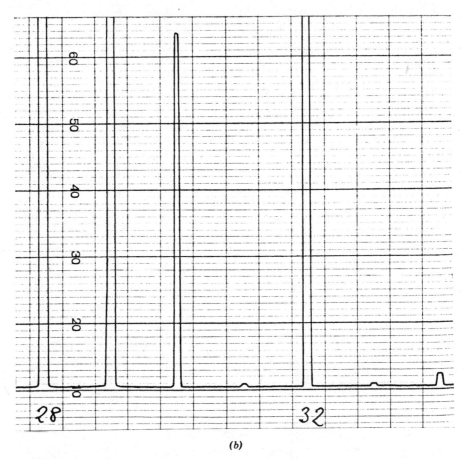

(b)

Figure 42 *(continued)* *(b)*   The same data as *(a)*, with a suppression lens used to eliminate the metastable ions. (Finnigan-MAT, GmbH)

cant information is lost. In fact, in this example the peaks remaining correspond
to an air leak, and the real sample data were suppressed!

To obtain useful metastable information, many manufacturers are now offering linked scan hardware. In this arrangement the accelerating voltage and the
magnetic fields are scanned together. In practice it is usually used on double-
focusing instruments, so that it is the ESA voltage that is scanned with the magnetic field. A number of modes are possible, notably B^2/E and B/E scans. These
give rise to parent and daughter ion displays. The peak shapes are not diffused
and can be detected readily by the data-acquisition system.

The data obtained are critically dependent on the scan conditions of the mass
spectrometer, and so the linked scan functions are usually under computer control. Combined with the display software this is a potentially powerful technique,
so much so in fact that relying on natural metastables may be an inefficient use

of instrument time. The introduction of a collision gas cell to promote decomposition of ions enhances the system's capability. Since the use of collision cells is not limited to sector instruments, they are discussed in more detail in the multistage mass analyzer section following a discussion of basic quadrupole mass spectrometry.

Quadrupole Mass Spectrometer

The quadrupole mass spectrometer was originally developed as a separator for uranium isotopes. Its use as a commercial enrichment system was not a viable proposition, but it was recognized that certain factors made it suitable for GC/MS work. Mechanically, the quadrupole is an extremely simple device, but a full understanding of quadrupole theory can be gained only by a detailed and complex mathematical study. By making a few assumptions and approximations it is possible to explain the principles of operation and to identify the critical instrument parameters.

The quadrupole consists of four rods arranged symmetrically as shown in Figure 43. Opposite rod pairs are electrically connected. To the x-axis rods is applied a *positive* voltage consisting of dc and rf terms.

$$V_x = +(V_{dc} + V_{rf} \cos \omega t) \tag{25}$$

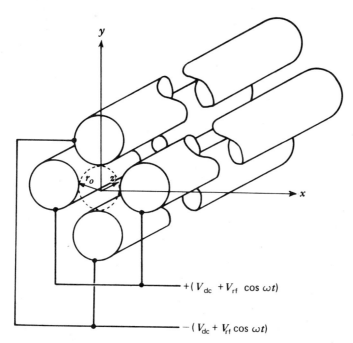

Figure 43. The quadrupole mass filter.

Note that the peak value of the rf voltage is greater than the dc voltage, so that during part of the rf cycle V_x will be negative. However, since the expression for V_x leads with a positive sign, the x-axis rods are often referred to as the *positive rod pair*. A similar voltage is applied to the y-axis rods.

$$V_y = -(V_{dc} + V_{rf} \cos \omega t) \tag{26}$$

The y-axis rods are sometimes known as the *negative rod pair*, although the voltage V_y will be positive during part of the rf cycle.

Examination of the voltages V_x and V_y shows that the rod pairs are excited with equal but opposite polarity voltages. Splitting out the rf and dc terms, we could consider the dc voltages as being equal and opposite in polarity and the rf voltages as 180° out of phase. Typical maximum values for the voltages are V_{dc}, 500 V and V_{rf}, 6 kV peak to peak. The rf frequency is held constant and has a typical value in the range 500 kHz to 3 MHz depending on the design. We can consider the rod pairs independently; that is, we can assume that the motion of the ions in the x direction is independent of the motion and position in the y direction and vice versa.

Consider the rod pair in the xz plane. There is no mass selection when only the dc voltage is applied (Fig. 44a). Positive ions will experience a force away from the rods toward the centerline of the assembly. Transmission of these ions through the quadrupole entrance and exit apertures for a given dc voltage will depend on the initial entrance conditions as well as on the mass. Applying both the dc and rf voltages (Fig. 44b) causes the ions to follow complex oscillatory trajectories. Some ions will pass through the exit aperture. For others the oscillations build up until the ion strikes the rod surface and loses its charge. Mathematical theory predicts a range of parameters represented by the coefficients a and q, where

$$a = \frac{8e/m \ V_{dc}}{r_o^2 \omega^2} \tag{27}$$

and

$$q = \frac{4e/m \ V_{rf}}{r_o^2 \omega^2} \tag{28}$$

which allow transmission.

Acceptable values for a and q are shown in the stability diagram of Figure 44c. Mathematical investigation of the trajectories shows that there is some degree of mass selection. Under normal mass spectrometer circumstances, the ion source will present a large number of ions of the same mass with a range of different entrance conditions. Not all of these ions will be able to pass through the rod space, even though they may be of the optimum mass for transmission. Hence, the performance of the *positive* rod pair can be represented by an attenuation diagram (Fig. 45). The optimum mass of maximum transmission occurs at the point of minimum attenuation. The attenuation rises more steeply for masses above the optimum than it does for masses below the optimum. The xz-plane *positive* rod pair can therefore be considered similar to a low-pass mass filter.

The *yz*-plane negative rod pair can be treated in a similar way. With only the negative dc rod voltage applied (Fig. 46*a*), positive ions would be strongly attracted to the nearest rod as soon as they moved off the quadrupole centerline.

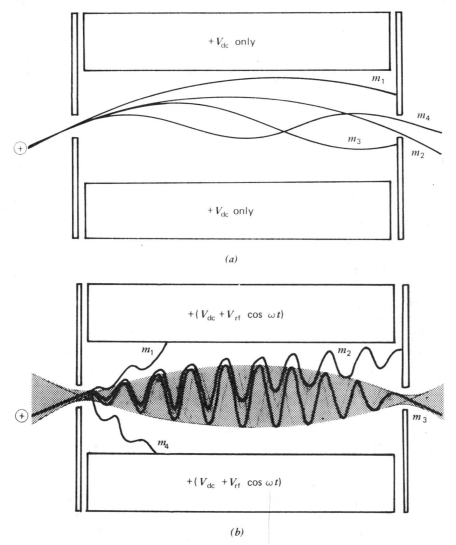

(a)

(b)

Figure 44. Characteristics of the *x*-axis quadrupole rod pair. (*a*) Ion trajectories with no rf voltage. (*b*) Ion trajectories with dc and rf voltages.

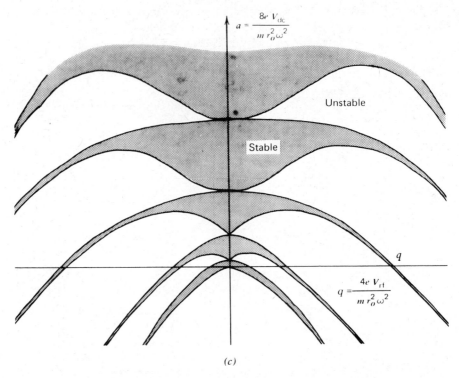

$$a = \frac{8e\,V_{dc}}{m\,r_o^2\,\omega^2}$$

Unstable

Stable

q

$$q = \frac{4e\,V_{rf}}{m\,r_o^2\,\omega^2}$$

(c)

Figure 44 (*continued*) (*c*)   Stable areas for parameters *a* and *q*.

There is no mass selection. Applying an rf voltage whose amplitude is greater than the dc voltage will cause the net rod voltage to be positive during part of the rf cycle. This has a stabilizing effect for some ions. The trajectories will be complex (Fig. 46*b*), but some ions will undergo oscillations of limited or bounded amplitude and will pass through the exit aperture. Reference to the mathematics again gives a diagram (Fig. 46*c*) representing acceptable ranges of the coefficients *a* and *q* for the *yz* rod pair that will give stable trajectories.

In terms of mass selection, the negative rod pair can also be represented by an attenuation diagram. Figure 45 shows that there is an optimum mass of maximum transmission, with attenuation rising more steeply for ions of lower mass than those of higher mass. The *yz*-plane negative rod pair can therefore be considered as a high-pass mass filter.

In practice, all the quadrupole rods are excited simultaneously. An ion entering the rod space is subjected to the transmission characteristics of both the positive and negative rod pairs at the same time. This can be represented by summing the separate attenuation curves to give the combined characteristic, which is also shown in Figure 45. It indicates that there is one mass of minimum attenuation. Ions of this mass will show maximum transmission, but it should be noted that not all will have the correct entrance conditions to successfully negoti-

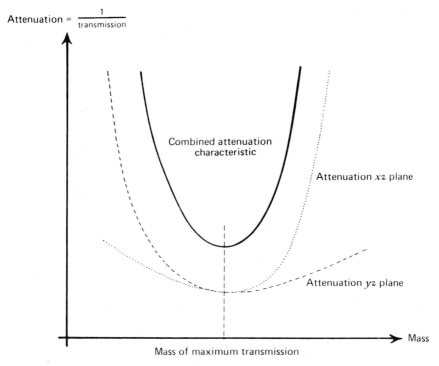

Attenuation = $\dfrac{1}{\text{transmission}}$

Combined attenuation characteristic

Attenuation xz plane

Attenuation yz plane

Mass

Mass of maximum transmission

Figure 45.   Transmission characteristics for the quadrupole mass filter.

ate the rf and dc fields. The attenuation at this optimum point represents the proportion of ions (assuming a statistically large number enter the quadrupole) that are rejected and does not refer to the transit time of those selected. The quadrupole produces no forces in the z direction, and once the ions enter the quadrupole space their z component of motion will remain unaltered. Hence, the quadrupole can be considered equivalent to a narrow-bandpass mass filter, the mass selected depending on the instrument parameters.

Similarly, we can predict the stability of an ion entering the quadrupole by combining both the xz and yz stability diagrams. This will give an infinite number of areas of overall stability (Fig. 47). These areas are mathematically sound for a hyperbolic field pattern, but most of them do not take into account the physical boundaries imposed by the rods themselves. The ion trajectories for some of these regions would show oscillations with amplitudes greater than the quadrupole's internal dimension r_o, the distance from the axis to the rod surface. These ions, along with those having unstable trajectories, will hit the rods and lose their charge.

The valid areas of stability are limited to the four approximately triangular areas arranged symmetrically about the origin of the aq axis. Consideration of the coefficients a and q shows that these four stable areas are mathematical

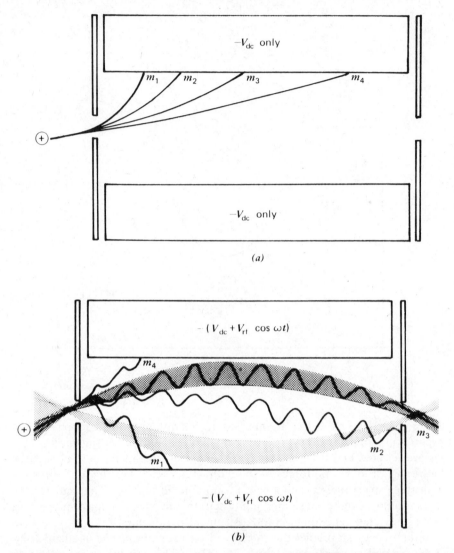

Figure 46. Characteristics of the y-axis quadropole rod pair. (*a*) Ion trajectories with no rf voltage. (*b*) Ion trajectories with dc and rf voltages.

manifestations of the same thing. We could, for instance, move from one zone to another by redefining the quadrupole axis or by interchanging the dc rod electrical connections or even by changing the rf phase. We need only consider one zone for a full understanding of the significance of the coefficients *a* and *q*. Figure 48 shows one of these regions in detail and is the diagram most often given to explain the principles of quadrupole mass spectrometer operation under dynamic conditions.

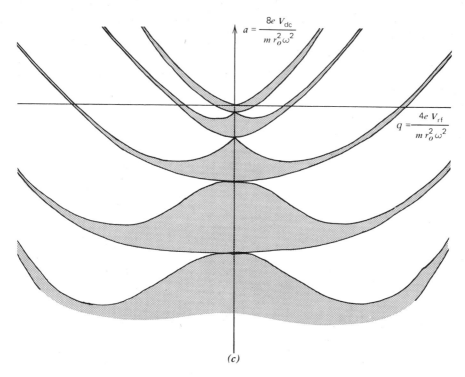

$$a = \frac{8e\,V_{dc}}{m\,r_o^2\,\omega^2}$$

$$q = \frac{4e\,V_{rf}}{m\,r_o^2\,\omega^2}$$

(c)

Figure 46 *(continued)* *(c)*   Stable values for parameters a and q

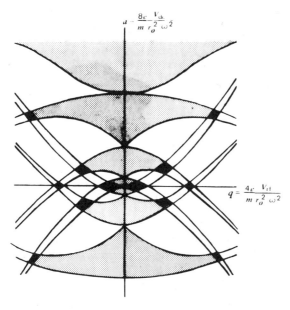

$$a = \frac{8e\,V_{dc}}{m\,r_o^2\,\omega^2}$$

$$q = \frac{4e\,V_{rf}}{m\,r_o^2\,\omega^2}$$

Figure 47.   Areas for parameters a and q that give three-dimensional mathematical stability for ions traversing the quadrupole.

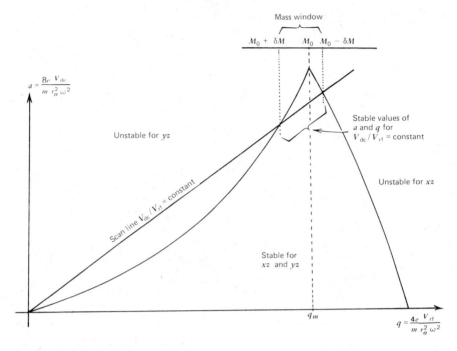

Figure 48. Stability diagram for practical quadrupole mass filter (enlarged portion of general a, q diagram).

The Quadrupole Under Practical Conditions. It is usual to operate the quadrupole with a fixed radio frequency. The frequency stability is very critical to performance and the rf voltage is usually derived from a very high-Q crystal-controlled power oscillator. The dimension r_o is similarly critical, and quadrupole rods are manufactured to very high dimensional standards. The arrangements for rod-to-rod fixing are equally important, and much consideration is given to mechanical and thermal stability. It is also usual to operate in such a manner as to maintain a fixed ratio between the rf amplitude and dc rod voltage. This can be represented by a straight line on the stability diagram (Fig. 48); this line is sometimes known as the *scan line*.

Under these conditions the values of stable a and q are limited to a small range. They depend only on the rf amplitude and the rf/dc amplitude ratio. The values of a and q for the mass of maximum transmission lie on a locus that passes through the apex of the triangular stability diagram. (This is an assumption taken from the mathematical study.) The value of q for an ion of this mass when the quadrupole is operated as described is given at the intersection of the locus and the scan line. Under normal operating conditions this gives a value of q approximately equal to the value at the triangle apex. For maximum transmission we have

$$\frac{e}{m} \, V_{rf} = \text{constant} \qquad (29)$$

giving a mass directly proportional to the rf amplitude. The quadrupole will therefore have a linear mass scale if the rf amplitude is scanned in a linear fashion.

The range of stable q values is set by the intersections of each side of the stability diagram with the scan line and is dependent on the slope of the line. This range will define upper and lower limits for the masses that can pass through the quadrupole. As a consequence the resolution is proportional to the rf/dc amplitude ratio. However, resolution is normally defined as the mass divided by the mass peak width (at some defined height). The expression for a shows that for a range of values above and below the value corresponding to maximum transmission the selected mass window will have a fixed width. Hence all mass peaks in a quadrupole spectrum will have the same width regardless of their mass. Resolution under the classical definition is therefore mass dependent. For a quadrupole mass spectrometer, resolution is sometimes quoted in terms of the mass that is selected. For example, the maximum resolution for a typical instrument may be quoted as $3M$, giving a value of 30 at mass 10, 300 at mass 100, and 3000 at mass 1000 u.

Quadrupole mass spectrometers are often accused of showing mass discrimination compared to sector instruments. Consider a sector instrument operated with a linear current scan at a resolution such that the high-mass ions of some standard compound are just resolved. Operating at a fixed resolution, this instrument will give fragment ion peaks at the lower end of the spectrum that are narrower than the molecular ion peak (Fig. 49a). A quadrupole set up for the same compound so that the high-mass ions are just resolved will give low-mass ions of the same mass peak width. The low mass ion peaks of the quadrupole instrument will therefore be proportionally bigger than those of the sector instrument spectrum.

Comparing the quadrupole and sector instrument spectra when normalizing to the low-mass ions (Fig. 49b) will indicate severe high-mass discrimination for the quadrupole. Normalizing to the high-mass ions for both spectra (Fig. 49c) implies severe low mass discrimination by the sector instrument. It is obvious that comparisons of this type are not valid. Also the pictures would be quite different for different scan conditions. Comparisons can be made only if total-ion and single-ion detection limits are compared directly. Specifications quoted for currently available instruments offer little evidence to support the mass discrimination claims. Having said this, there are some factors that will cause the quadrupole to show poorer high-mass to low-mass ratios than might have been expected.

For a given mass separation it is necessary for the ions to undergo an optimum number of rf cycles as they traverse the quadrupole. There is no acceleration in the z direction, so the ion velocity through the quadrupole will be the

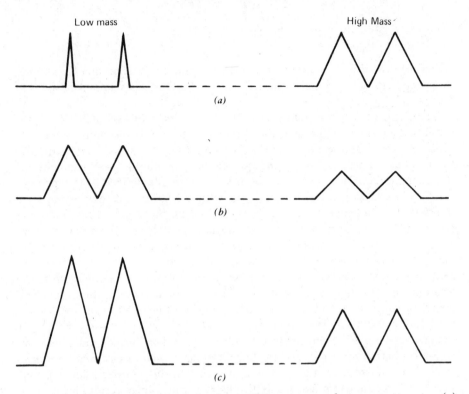

Figure 49. Apparent mass discrimination in different types of mass spectrometer. (*a*) Sector instrument spectrum; upper mass equivalent to operating resolution. (*b*) Quadrupole spectrum of same compound normalized to base peaks. (*c*) Quadrupole spectrum normalized to high-mass peaks.

same as the entrance velocity. This depends on the energy equation of the ion source, potential energy lost being equal to the kinetic energy gained.

$$eV = \tfrac{1}{2}mv^2 \tag{30}$$

where m is the ion mass, v is the entrance velocity, V is the ion energy voltage, and e is the charge on the ion. Consequently, the ion velocity is mass dependent, and not all masses will undergo the optimum number of rf cycles for separation. The effective transmission may well be reduced, giving some mass discrimination. This effect can be overcome by programming the ion energy voltage upwards as the mass is scanned, giving a constant ion velocity through the quadrupole. Ions of different mass will then each experience the optimum number of rf cycles for proper resolution and transmission.

Quadrupole Rod Shape. There has been much discussion in recent years about the shape of the quadrupole rods. The mathematics requires a hyperbolic field pattern, which in theory is achieved by four symmetrically arranged hyperbolic surfaces. This, of course, is an impossible structure to use in the real world. Adjacent rods must be excited with equal but opposite voltages. Hyperbolic surfaces approach asymptotically, touching at infinity. In practice the clearance is very small after only a short distance. The hyperbolic surface must be truncated to prevent shorting and arc-over of the rf and dc voltages. Hence, the field pattern for both circular and hyperbolic section rods is only an approximation to the theory.

Manufacturers have used various shapes and modifications in an attempt to reach the ideal. For instance, Hewlett-Packard once made improvements by adding trimmer stubs to each side of their round rods, but the duodecapole, as it was called, was then a complicated device to make and set up. Manufacturers will generally choose a rod shape that leads to the easiest solution of the field approximation problem consistent with the constraints of manufacture, assembly, and testing. With the different manufacturing technologies, experience, and skills available to them, it is not surprising that companies choose different approaches to quadrupole design. Figure 50 shows some of the more recent arrangements.

Quadrupole performance will depend not only on the closeness of the field approximation but also on other factors such as end fringe fields, surface quality, and electronics stability. The resultant spectral peak shape is far more critical than the actual quadrupole rod shape and will depend not only on the parameters just mentioned but also on how the instrument is used and maintained.

Quadrupole Scan Characteristics. Whatever shape or size the rods are, the quadrupole is essentially an electrostatic device. It has a very low inductance and a relatively low capacitance and can therefore be scanned at high rates. That is to say, the amplitudes of the rf and dc voltages can be changed extremely quickly; in most cases a few milliseconds is sufficient for a change from minimum to maximum values. The device is therefore suitable for both fast scanning work and selected ion monitoring. In the latter mode the rf and dc voltages are rapidly switched among a number of values corresponding to the masses of interest.

The quadrupole will also act equally well on ions of either positive or negative charge. This is in effect equivalent to simply changing the rod polarity. The quadrupole is suitable for simultaneous positive and negative ion work providing the problems of detection caused by the relatively low ion energy can be overcome. Suitable detection arrangements are discussed in following sections.

Multistage Mass Analyzers

Quadrupole instruments are continuous mass-focusing devices and do not show metastable peaks in their spectra. However, a development of the instrument using three quadrupoles in series has led to a new area of quadrupole mass spectrometry. The triple quadrupole arrangement (Figs. 51 and 52) provides an ini-

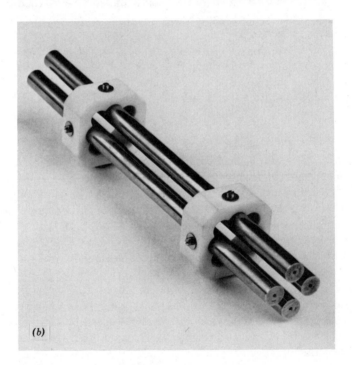

Figure 50. Quadrupole rod assemblies. (*a*) Hewlett-Packard, hyperbolic section. (*b*) Finnigan-MAT 4000 series, round section.

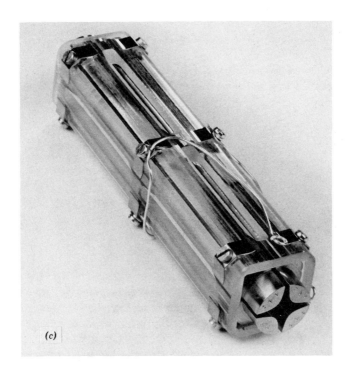

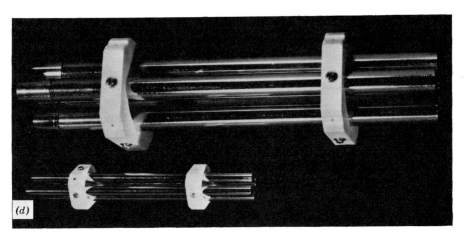

Figure 50 (*continued*)   (*c*) Finnigan-MAT 4500 series, hyperbolic section. (*d*) Nermag, large-diameter round section.

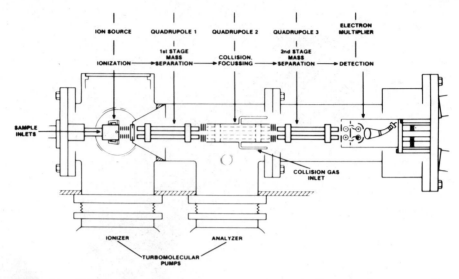

Figure 51. Triple-stage quadrupole mass spectrometer, diagramatic cross section. (Copyright © 1982 Finnigan Corporation. All rights reserved.)

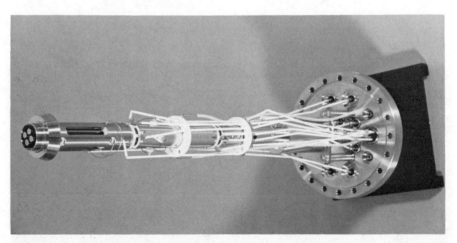

Figure 52. Triple-stage quadrupole assembly. (Copyright © 1982 Finnigan Corporation. All rights reserved.)

tial mass-selecting filter. This is followed by a second quadrupole, which acts as an ion-retaining reaction cell where collision-induced fragmentation can occur. The third quadrupole acts as a mass filter to monitor the resultant daughter ions.

This arrangement compensates quadrupole instruments to some degree for the lack of metastable response. In fact, metastables are artificially induced when the ion beam from the first quadrupole collides with gas molecules in the second. Linked scanning arrangements between the quadrupoles can be set to collect specific data such as data associated with a given fragment loss. A lot of structural information about a compound can be obtained; the parent/daughter map display of Figure 53 is typical.

Collision-induced decomposition is not restricted to the quadrupole arrangements discussed above. Collision cells, small differentially pumped zones, have been included in the ion flight path of several sector instrument designs. These are filled with low-pressure gas, the molecules of which collide with the ion beam, causing further fragmentation. The arrangement is available as an option from a number of manufacturers of double-focusing instruments. The collision cell is usually placed between the ion source and the electrostatic analyzer. The ESA and the subsequent magnetic sector are then operated in a linked scan arrangement to monitor the collision-induced ions.

For serious mass/ion kinetic energy studies, a reverse geometry is recommended. That is, the magnetic sector is placed before the ESA. This type of instrument usually has a number of collision cells before and after the magnet and is generally purchased specifically for MIKES* work; it is mentioned here for general information only.

The link scanning techniques required for both the multistage quadrupole and sector instruments rely heavily on computer control. This is also true of the data acquisition and presentation. The integration of hardware and software in scientific instrumentation is advancing steadily, and it would be reasonable to expect further developments in multistage analyzers as a result. Multistage instruments offer a number of interesting possibilities, one of which is the removal of the necessity to separate compounds using the gas chromatograph! However, the high cost of the necessary hardware, the problems associated with dirty samples and source and collision cell contamination may delay the demise of the combined GC/MS system for some time yet.

SIGNAL DETECTORS

The ion current flowing through the mass spectrometer in a GC/MS system is extremely small, but it may change rapidly over a wide range. To detect this current requires electronic circuits of high signal gain, wide dynamic range, and good frequency response. At these low levels, typically in the range 10^{-14} to 10^{-11} A, noise from extraneous sources is a very significant factor, and great care is taken with amplifier design. A current of 10^{-15} A, equivalent to about 6000 singly charged ions arriving at the detector per second, is a practical detection limit for normal GC/MS systems. Below this level, the signal fluctuation caused by the statistical spread of ions with time may well be greater than inherent noise of the amplifier circuits.

*Mass analyzed Ion Kinetic Energy Spectrometry.

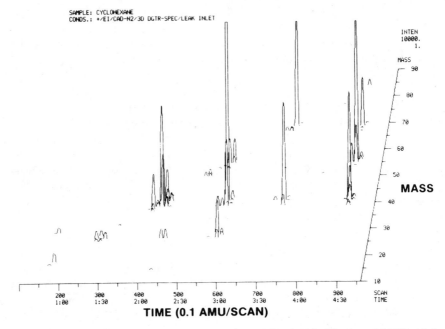

SAMPLE: CYCLOHEXANE
CONDS.: +/EI/CAD-N2/30 DGTR-SPEC/LEAK INLET

TIME (0.1 AMU/SCAN)

Figure 53. Typical parent/daughter ion map obtained with Finnigan-MAT's triple-stage quadrupole instrument. The parent ions for cyclohexane lie on an axis at about 45° on the graph at the top of scans showing resultant daughters from collision-activated decompositions. (Copyright © 1982 Finnigan Corporation. All rights reserved.)

Even above this limiting level the signals must be amplified significantly before the usual signal-processing and data-handling techniques can be applied. The amplification is usually achieved with an electron multiplier mounted within the mass spectrometer vacuum envelope. In many cases it is the high voltage on the electron multiplier with the attendant danger of flashover which sets the pressure limit for the GC/MS system. Multipliers provide a signal gain in the range 10^4 to 10^6. The current flowing out from them is still extremely small and is usually further amplified by high-input impedance amplifiers called electrometers.

The purpose of the electrometer is to provide impedance matching between the multiplier and the signal-processing circuitry. The electron multiplier is essentially a current amplifier, whereas most of the signal processing is done by voltage-dependent devices. For the multiplier to produce a relatively high signal voltage, its output must be coupled to a very high value resistor. The input impedance of measuring circuits must consequently be higher if current loading and leakage are to be minimized. The details of how electrometers work, and indeed which type of electrometer should be used, are beyond the scope of this text. Sufficient to say, the electrometer or preamplifier, as it is sometimes called, must provide a very high input impedance, a current-to-voltage conversion, and a relatively low output impedance to drive the signal-processing electronics.

The preamplifier is usually mounted very close to the electron multiplier to minimize stray pickup and noise. Both the multiplier and preamplifier are usually mounted so as to minimize the effects of mechanical vibration which can produce noticeable microphonic signals. The multiplier, being the first amplification stage in the signal-processing chain, is the most critical component as far as noise is concerned. A very high signal-to-noise performance is achieved with a number of multiplier designs, the principal types in use being box-and-grid, venetian blind, and continuous dynode multipliers.

Electron Multipliers

Box-and-Grid Multiplier. The ion beam from the mass spectrometer is accelerated toward a fine mesh or grid that is held at a high negative potential. Most of the ions pass through the grid to strike a copper-beryllium surface held at the same potential. This surface is very emissive and will release three or four electrons for each impinging ion. The electrons are accelerated out of the boxlike area toward a second grid-and-box arrangement whose potential is held at about 200 V below the first. The electrons striking the second copper-beryllium surface cause further electron emission, and the process is repeated typically 12 to 16 times, as shown in Figure 54. Each stage is at a slightly lower potential defined

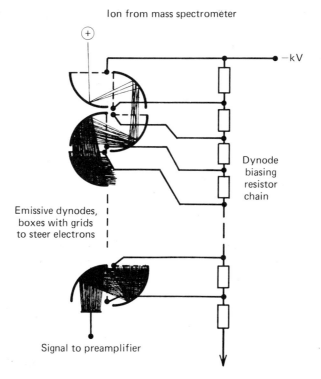

Figure 54. Box-and-grid multiplier—principles of operation.

by a chain of resistors, so the total applied voltage must be in the range of 2.5 to 3.5 kV.

The overall gain is very dependent on the condition of the surfaces as well as on the applied voltage. At 3 kV a gain of 10^6 is typical for a multiplier in good condition. The efficiency falls off as the surfaces become dirty due to contamination by water vapor, and the gain can fall dramatically after only a short exposure to the atmosphere. Fortunately, the procedure to rejuvenate the surfaces is straightforward and usually successful.

Venetian Blind Multiplier.   Like the box-and-grid multiplier, the venetian blind multiplier relies on the emissive properties of a copper-beryllium surface. Curved plates are arranged like the slats of a venetian blind (Fig. 55) with a potential of about 200 V between plates. Ions strike the first plate, causing electron emission. The electrons skip down the plates, causing a cascade of further emission from each subsequent stage. Typically, each stage contributes a factor of 2 or 3 to the gain, and so about 16 stages are needed. Increasing the number of stages brings a diminishing return. Above a gain of about 10^6, or less if the input current is high, the output stages will saturate. The biasing resistors limit the level of output current and so prevent adverse ion burns on the active surfaces. The overall performance and limitations of the venetian blind multiplier are very similar to those of the box-and-grid multiplier. Both can be rejuvenated by treating the copper-beryllium surfaces according to a procedure discussed in the routine maintenance section (see later).

Continuous Dynode Multiplier.   A high negative voltage, 1800 V typically, is applied across a glass trumpet whose inner surface is coated with tin oxide. The coating serves both as the emitting surface and as the potential dropper. It is equivalent to a continuous thin-film metal oxide resistor, and a potential gradient is developed down the envelope. An ion striking the surface will cause several

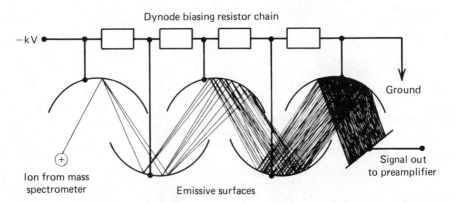

Figure 55.   Venetian blind multiplier—principles of operation.

electrons to be emitted. These will skip down the surface, causing further electron emission. A gain of 10^5 to 10^6 is typically achieved with 1800 to 2000 V. To prevent loss of electrons out of the mouth of the trumpet and to define the entrance potential, a fine mesh is usually fitted in electrical contact with the high voltage supply across the top of the multiplier. The arrangement is shown in Figure 56.

The continuous dynode multiplier, like other types of multipliers is extremely susceptible to surface contamination, and cleaning often gives a considerable gain improvement. Surface damage due to ion beam and electron bombardment as well as contamination is not so easily restored. A number of rejuvenating procedures have been proposed, but the success rate is not high. Fortunately, the surface is not sensitive to water vapor, air, or the most common CI reagent gases. The multiplier can therefore be used with confidence in even the most severe GC/MS environments. It can be removed for cleaning or exposed to the atmosphere during system shutdown without sudden loss of sensitivity.

X-Ray Shield

When ions strike the rods in a quadrupole system, soft x-rays may be generated. X-rays impinging on the signal-detecting multiplier may cause electron emission from the surface by photoelectric action. The current due to the x-rays can completely mask the ion beam signal. Even in less severe cases the effect causes detection problems. As the partial pressure of unselected ions rises (that is, ions at masses not scanned by the quadrupole, the x-ray current produced by collisions with the rods increases. This causes a rising tide of signal and noise to sweep over the displayed part of the mass spectrum. The effect is greatly reduced when the

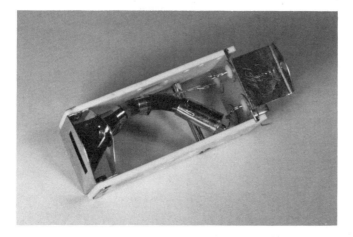

Figure 56. Continuous dynode electon multiplier.

quadrupole is tuned to the more intense ions, as there will then be fewer x-rays and more detected signal.

In addition to any x-rays produced, ion-rod collisions generate significant numbers of excited neutral molecules. These too can give rise to multiplier output signals if they impinge on the active multiplier surfaces. The effect is greater with increasing ion current and with increasing amounts of gas entering the quadrupole structure. For example, the multiplier noise level can increase considerably in the CI mode with a large air leak or with any large ion signal that is not being mass selected by the mass spectrometer.

These effects are obviously quite intolerable in a GC/MS system where the partial pressures of different masses are changing as compounds elute. To overcome this problem the multiplier must be mounted behind an x-ray shield as shown in Figure 57. This arrangement has no line of sight between the rods and the multiplier so that x-rays are prevented from hitting the surfaces. The x-ray shield also serves as a baffle to keep out excited neutral molecules. Note that the multiplier illustrated is a continuous dynode type, but the effect is common to all multiplier designs. The problem now is to get the ions past the x-ray shield into the multiplier.

Positive ions are attracted strongly into the multiplier because of the high negative voltage at its entrance. Even so, the ions must be pulled through a defined area without striking the surrounding shields. For the heavier ions the minimum potential gradient necessary to overcome the ion's inertia and pull it around the x-ray shield is greater than for ions of lower mass. This sets a mass-dependent minimum limit on the multiplier operating voltage. Typically the

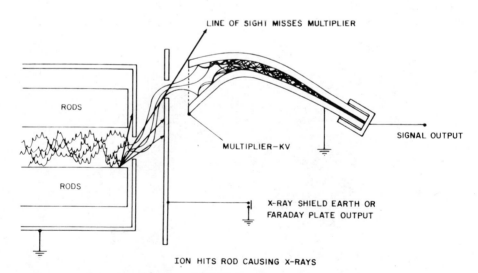

Figure 57. X-Ray shield fitted to a quadrupole mass spectrometer. (Copyright © 1982 Finnigan Corporation. All rights reserved.)

multiplier must be operated above 1700 V for an 800-u instrument if all masses are to be satisfactorily presented to the multiplier.

Magnetic sector instruments are not subject to this problem, although x-rays could be produced by ions striking the vacuum enclosure. The collector slit and the curved flight tube serve as an integral shield that allows few x-rays through to the multiplier. It should be pointed out that in GC/MS systems the x-ray energy is low and the x-rays will not normally penetrate the mass spectrometer vacuum envelope. The system operators are therefore protected from harmful radiation.

The x-ray shield will charge up and deflect the ion beam if it is not adequately grounded. Often the earth connection is made outside the vacuum chamber with a shorting connector. If the multiplier is switched off and the shorting plug is removed, the plate can be used as an ion collector. An amplifier can be connected to the shield and the ion signal monitored. This feature is useful for measuring the signal level before multiplication so that the multiplier gain can be assessed.

Total Ionization Monitors

It is often useful to know that a certain level of ionization is being achieved in the ion source even though all of it may not be directed to the signal detector. Furthermore, the total ionization signal is analogous to the outputs of other GC detectors so that the output is in effect a gas chromatogram. The information is obtained from a total ionization monitor, TIM. In sector instruments the total ionization monitor usually consists of a plate with a rectangular hole mounted just after the source exit slit. The plate is designed so that most of the ion beam passes through it regardless of the source slit setting. About 10% of the ion beam, that is, 5% at either end, is intercepted by the TIM plate. The current flowing to the TIM plate is measured and the total ionization can then be calculated.

The signal response requirements and detection sensitivity are not as severe as in the case of output signal measurement. Consequently, total ionization monitors can be made using a very high-gain narrow-bandwidth electrometer amplifier. It is not uncommon for there to be several monitor points along the ion path on large sector instruments. Often a single amplifier with one or more electrometer heads is used for measurement. The heads may be selected by the operator to obtain the ion beam transmission characteristics.

Quadrupole systems are not normally operated with TIMs in this way. It is not practical to intercept a fixed percentage of the ion beam without upsetting the action of the source-focusing plates or the quadrupole rods. Total ionization must be assessed by integrating the detected mass spectrum, usually with an electronic integrating circuit. Of course, only the scanned mass spectrum will be integrated, which may not completely reflect the total ionization from the source. On the other hand, an advantage of this approach, which is not limited just to quadrupoles, is that the spectrum may be gated electronically so that only a selected mass range is integrated even though a larger range is scanned. This technique allows the operator to observe low-intensity mass ions in the presence of a high background without the background appearing on the TIM output. This is very useful in suppressing carrier gas, reagent gas, and solvent background signals from the recorded chromatogram. With computer manipulation of collected GC/MS data, this technique can be easily carried further so that limited mass range chromatograms can easily be displayed.

Negative Ion Detectors

Sector Instruments.   The high negative voltage at the entrance of the electron multiplier is fine for attracting positive ions but will repel negative ions. This is not normally a problem with sector instruments operating in a negative-ion mode, as the ions usually have sufficient energy to overcome the multiplier entrance potential. For example, a typical system with an ion source operating at 5 kV and a multiplier voltage of 3 kV will produce ions with 2 keV of excess energy. Hence, these ions will be able to enter the multiplier in the normal way. Of course, if the system is to be operated in the voltage-scanning or voltage-switching modes, this excess energy will be reduced as the accelerating voltage is scanned or stepped downwards. Since the mass selected is inversely proportional to the accelerating voltage, the system will show a lower working mass range for negative ions than for positive ions. It is not usual to operate electron multipliers much above 3 kV so that sector instruments operating with source potentials well above 5 kV are not subjected to these negative-ion mass limits.

Quadrupole Instruments.   These normally operate with ion energies in the range of 2 to 20 eV. This energy is quite insufficient for a negative ion to enter the normal electron multiplier arrangement. One way to overcome this problem is to offset the multiplier voltages so that the potential presented to the ions is about 2 kV positive. Unfortunately, this means that the signal output end of the multiplier must also be raised in potential. The actual voltage will depend on the type of multiplier used and is typically 4 kV for a continuous dynode multiplier and 5 kV for a box-and-grid type. This high output voltage cannot be connected directly to the signal-processing electronics, so some form of isolation amplifier must be used. The design of such circuits is not a straightforward matter, and achieving a satisfactory signal gain and noise level is no mean feat and is always expensive. Furthermore, if simultaneous, or at least rapidly switched, positive and negative ion collection is required, a second multiplier must be used. This is because it is not practical to change the multiplier voltages on a single multiplier quickly enough without generating large signal transients on the amplifier circuits.

An alternative method of positive and negative ion collection using only one multiplier depends on the action of two conversion dynodes placed before the multiplier, as shown in Figure 58. High-energy ions striking a suitably prepared

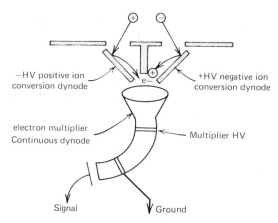

Figure 58. Conversion dynodes added to a continuous dynode electron multiplier for positive and negative ion detection. (Copyright © 1982, Finnigan Corporation. All rights reserved.)

surface will cause secondary emission of electrons and protons as well as other forms of radiation. By arranging two conversion dynodes, one at a high positive potential, say $+3$ kV, and one at a high negative potential, say -3 kV, it is possible to separate the two ion beams leaving the quadrupole. The positive ions are pulled to one side of the x-ray shield to strike the negatively charged conversion dynode. Any secondary electrons emitted will be repelled away from the dynode toward the multiplier entrance, which is at the relatively lower potential of about -1500 V.

Negative ions are drawn through the other hole in the x-ray shield to strike the positively charged dynode. Any positive ions released will be repelled from the dynode and strongly attracted into the multiplier. In both cases, photoemission from the dynodes to the multiplier is also an advantage, as these rays arise from the selected ions. The x-ray shield above the dynodes prevents unwanted radiation from unselected ions entering the multiplier. The dynodes are designed and arranged so that the relevant secondary ions are accelerated into the multiplier rather than to the opposite dynode.

The system illustrated employs a continuous dynode electron multiplier, but other types could be used. If the multiplier voltage is increased, then the conversion dynode voltage must be increased in proportion. To detect and separate the two signals requires some electronic gating circuits in the amplifier output stages. These must be switched in synchronism with the ion source electronics so that first positive ions are produced and detected, followed by the negative-ion cycle. Alternate scan switching and continuous modulation of the ion beam have both been tried with equal success. The electronics complexity and cost increases with switching speed, so the alternate scan method is usually used.

The conversion dynode system gives a significant improvement in signal-to-noise ratios compared to a multiplier alone. This improvement is not just limited to the negative-ion beam but also applies to the positive-ion signal. Its application is therefore not just limited to negative-ion work. The noise improvement arises from the additional gain of the conversion dynodes at the front end of the amplifier chain. In general terms, noise levels are most reduced when the gain is applied to a system as close as possible to the signal source. Not only is gain improved at the front end in this arrangement, but also the multiplier is usually operated at lower voltage. This improves its working life and its general noise level by reducing field emission from the active surfaces. For these reasons, conversion dynodes multipliers are quite often fitted even on single-polarity ion-detection systems.

CHAPTER FOUR

Gas Chromatography

Some of the earliest chromatographic work involved the separation of plant pigments into discrete colored bands. The term *chromatography* (colored writing) was in this case literally true, and it has since become the generic name for a wide range of separation sciences. It is the purpose of the chromatograph to separate out the components of what may be a complex chemical mixture. In the context of GC/MS we are concerned mainly with gas-liquid chromatography although gas-solid chromatographic techniques are sometimes used. A typical arrangement (Fig. 59) consists of a heated injector assembly connected to a column mounted in a temperature-controlled oven. A carrier gas is passed via the injector through the column and into the sample detector.

A small quantity of the chemical mixture to be analyzed is introduced into the gas stream by injection through the injector septum with a syringe. The various components of the sample travel through the column at different rates, some being retained longer than others, and are then presented to the detector in turn. In our case the detector is the mass spectrometer, which is usually preceded by a separator, a sample-enrichment device.

GAS CHROMATOGRAPHY TERMS AND ABBREVIATIONS

Bleed	Outgassed or vaporized material from either the septum or column packing. Often vaporized liquid phase.
Capillary column	General description of small-diameter columns without packing material.
Carrier	The gas passing through the column.
Dead volume	Part of the gas chromatograph or GC/MS interface with a relatively large volume (maybe only a fraction of a milliliter) where gas velocity may be reduced and with turbulence remixing of the sample may occur.

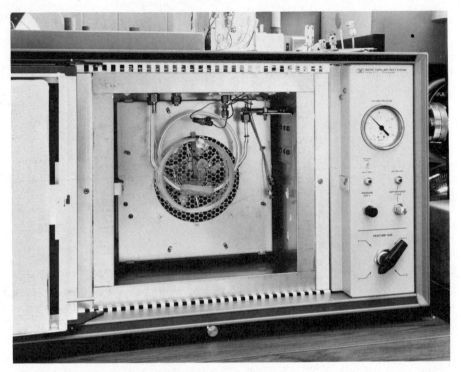

Figure 59. A gas chromatograph with both packed and capillary columns installed. (Hewlett-Packard Corporation)

Detector	Device for indicating the presence of compounds eluting from the gas chromatograph.
ECD	Electron-capture detector.
FID	Flame ionization detector.
Gas phase	Another name for the carrier gas.
GC	Gas chromatograph or gas chromatography, depending on context. Often used instead of GLC, gas-liquid chromatography.
Grob injector	Type of injector used for splitless capillary injection methods.
HETP	Height equivalent of a theoretical plate. Equivalent length of column required to produce one theoretical plate.
Injector	Point of introduction of the sample to the column.
Liquid phase	Liquid bonded to the support that plays an active part in sample retention in the columns.

Packed column	Column filled with a packing of liquid-coated support material.
Packing	Inert material coated with liquid phase, used to fill packed columns.
Peak	Output of a detector in response to a sample compound eluting from the column. Throughout this text the phrase *gas chromatograph peak* will be used to differentiate gas chromatograph detector responses from other peaks in GC/MS such as mass spectral peaks.
Peak broadening	Fault condition when gas chromatograph peaks are wider than normally expected.
Phase	The gas, liquid, or solid part of the gas chromatograph column arrangement, depending on context. Very often the term *phase* is used for liquid phase.
Retention time	Time taken, measured from injection, for a sample compound to elute from the gas chromatograph. Corrected retention time measures the retention time relative to the solvent peak to compensate for the gas flow or holdup time. This is usually significant only in capillary column work.
Resolution	Degree of separation between adjacent gas chromatograph peaks.
Sample	Material injected into the gas chromatograph. It may be a pure compound or many compounds and is usually, but not always, dissolved in a solvent.
SCOT column	Support-coated open tubular capillary column.
Septum	Soft material through which sample is injected.
Solid phase	Support material.
Splitter	Device for dividing gas and/or sample flow. Used in capillary injectors and GC/MS interfaces.
Splitting	Fault condition giving ragged or separate gas chromatograph peaks of the same compound.
Stationary phase	Liquid phase.
Support	Inert material onto which the liquid phase is bonded. Used as packing in gas chromatography columns. In the case of capillary columns, the support may be the inner wall of the column itself.
Tailing	A detector response that shows a nonsymmetrical gas chromatograph peak with a sharp rise but a long slow fallback to the baseline.
Theoretical plate	Hypothetical parameter representative of column efficiency.
WALLCOT column	Wall-coated open tubular capillary column.

THEORY OF GAS-LIQUID CHROMATOGRAPHY

Consider the hypothetical model of a gas-liquid chromatograph shown in Figure 60. This consists of a large number of reservoirs that each hold a relatively non-volatile liquid. Passing over the liquid is a gas stream into which is injected a discrete amount of a sample mixture. The sample will travel in the gas phase above the liquid phase and will start to dissolve in the liquid. The amount at any one time that is dissolved will depend on the nature of the sample and the liquid, the partial pressure of the various components in the sample, and the system temperature. The partial pressure of each component will vary as the gas flow sweeps the sample along. The dissolved components will reenter the gas stream as their partial pressure above the liquid phase drops. The sample will migrate down the system, different components traveling at different speeds. All will travel at a slower rate than the gas, and each will show a different retention time.

There are a number of practical realizations of the theoretical model. The most common consists of a glass or metal column filled with a packing material made from uniformly graded particles of an inert support material. Bonded or absorbed onto the surface of the support material is a liquid phase. The coating thickness or loading is very small, typically a few percent, so that the packing does not feel wet. Nevertheless, the coating behaves as a liquid and can dissolve and react with any chemicals passing over it. The coated support is tightly packed into the column and is held in place at each end by a piece of glass wool.

An inert gas is passed through the column. The gas connection is made at the head of the column, usually through an injector arrangement. The injector has a septum through which the sample can be introduced. The gas flow through the injector is arranged so that it sweeps the underside of the septum, carrying all the sample onto the column. The injector area is usually heated independently to a temperature significantly above that of the column to prevent sample conden-sation. The column is also heated and the temperature very carefully controlled.

During the chromatographic run, the column oven temperature may be held constant (isothermal operation) or varied (temperature programming). The ac-

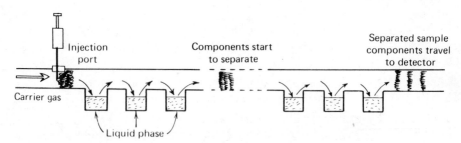

Figure 60.   Hypothetical model of a gas-liquid chromatograph.

tual temperature and the mode employed will depend on the required analysis. Higher temperatures give shorter retention times for a given sample compound. Temperature programming speeds up the analysis time when mixtures of compounds with significantly different molecular weights are tested. In all cases care should be taken not to exceed the upper operating temperature limit for the column or it may be destroyed. Even at temperatures somewhat below the upper limit, loss of phase from the support material—that is, column bleed—may be significant. Also, it is worth noting that there is a lower temperature limit for the column. Below this temperature the liquid phase may solidify and become ineffective.

Column efficiency is measured in terms of the gas chromatography peak width. Compounds do not elute from the column instantaneously. At the end of the retention period the rate of elution begins to rise, reaches a maximum, and then falls again. For packed columns this peak width is typically 15 to 20 s. The narrower the peak, the more efficient the column. The sharpness is quantified by considering the number of theoretical plates in the column. The plate concept comes from the early days of chromatography work when distillation techniques were the normal method for separating compounds and, although theoretical in nature, is valid when evaluating and comparing column performance.

Assuming triangular gas chromatography peak shapes, that is, discounting peak lead and tail, the number of theoretical plates N is given by

$$N = 16\left(\frac{X}{Y}\right)^2 \tag{30}$$

where X is the retention time and Y is the baseline peak width also expressed as a time. It is normal to discuss the number of theoretical plates per unit length of column or the height equivalent to a theoretical plate, HETP, where

$$\text{HETP} = L/N \tag{31}$$

where L is the column length.

Typical values for packed columns are 1000 to 2000 plates per meter. Capillary columns give higher figures, typically 4000 to 5000 plates per meter. Figure 61 illustrates how the number of theoretical plates can be calculated for packed and capillary columns. Note that the capillary calculation is corrected for the gas-flow delay through the column. The retention time is computed from the solvent peak, not from the injection point. The solvent elution time for packed columns is typically 10 to 15 s and is not significant except for calculations based on gas chromatography peaks with very short retention times. In this case a corrected retention time should be used as the basis for the calculations. The peak width is determined by drawing tangents at the point of inflection on the GC peak down to the baseline. This is valid only if the peak is symmetrical. Gas chromatography peaks showing tailing or other distortions should not be used

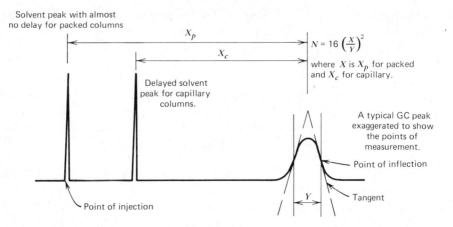

Figure 61. Calculation of number of theoretical plates for a gas chromatography column.

for HETP calculation but nevertheless should not be ignored as they indicate the general quality of the column.

The ability of a column to separate components in a sample mixture depends not only on the number of theoretical plates but also on the solvent efficiency of the phase being used. Columns of different phases but with the same column efficiency will have different separating powers for a given sample. The solvent efficiency is related to the affinity of the liquid phase for the sample components and is often expressed in terms of the polarity of the liquid-phase material.

Once a type of column packing has been chosen for an analysis, the column efficiency is the most important parameter not only in separation power but also in detection limits. The relative sample concentration is greater in narrow (sharp) gas chromatography peaks than in wide ones. Narrow peaks are achieved by using columns of high efficiency and analytical methods giving retention times as short as possible consistent with the required separation from interfering compounds. Short analysis times are also generally desirable, as this allows greater sample throughput, maximizing the utility of the GC/MS system.

GAS CHROMATOGRAPHY HARDWARE

The essential hardware units of a gas chromatography system are injector, column, and detector. Equally important are such things as syringes, gas lines, filters, and fittings. Detailed discussion of various aspects of this hardware with regard to successful operation of a GC/MS instrument is to be found in later sections. An outline of the main types of gas chromatography hardware commonly used with GC/MS systems is presented in the following paragraphs.

Columns

The GC column is the heart of a GC/MS system. Without its separating power the mass spectral data would be impossible to interpret. The degree of resolution required for adequate separation will depend on the complexity of the sample being analyzed. Simple mixtures are best suited to packed columns, which can give relatively short analysis times. Complex mixtures usually require the higher resolution of capillary columns with correspondingly longer analysis times.

Packed Columns. The typical length of a packed column is 2 m. In special cases, especially with a single compound or simple mixture, very short columns may be used. These are particularly useful for delicate compounds where the analyst is anxious to get the sample into the mass spectrometer as soon as possible. On the other hand, increasing the column length increases its separating power. Unfortunately, the law of diminishing returns sets in. The inside diameter of a packed column is typically 2 mm. The column is filled with fine particles of packing material coated with the liquid phase. There are many routes for the gas to flow around this packing, and as the column length is increased the likelihood of sample remixing and gas chromatography peak broadening increases. This *multipath effect* sets an upper limit on column length of about 3 to 4 m. In most practical gas chromatographs, 3 m is also about the maximum length of coiled glass tube that can be accommodated. Coiling the glass tube too tightly increases the multipath effect by providing a shorter "inside lane".

The outside diameter of packed columns is typically 1/4 in., although 6 mm is now common in some parts of the world. It is normal practice to get columns made by a local glass worker and then fill them with the required packing. In this case, or even when purchased from the GC/MS manufacturer or his agents, always check the column outside diameter and use fittings and ferrules of the correct size. Although there is not much difference between 6 mm and 1/4 in., it is more than enough to give a lot of sealing and dead volume problems. This advice is also valid for metal columns. There is a great temptation to overtighten metal fittings, especially as they are so much stronger than glass. The result is often a damaged fitting to the injector or the GC/MS interface. The use of metal ferrules is also of doubtful value. Inevitably these swage onto the metal column and cannot be replaced. In general terms, metal ferrules should be used only on permanent or semipermanent joints. It is unlikely that a GC column connection would fall into these categories, so a softer ferrule is recommended.

Capillary Columns. This type of column has no packing material. The liquid phase is bonded either directly to the column walls (WALLCOT) or to a support material coating the inner wall surface (SCOT). There is no multipath effect (peak broadening due to the tortuous paths of the gas through the packing), so column efficiency is improved. Four or five thousand theoretical plates per meter are typical. Furthermore, without the multipath problem it is feasible to extend the column length; 25- or even 50-m columns are normal.

For there to be effective interaction between the gas and liquid phases, the internal diameter needs to be small, typically 0.1 mm. Gas flow rates are therefore low, 0.1 to 1 mL/min, and connection can often be made directly to the mass spectrometer system without the need of sample–carrier gas separators. This is not only a simple and practical arrangement but also one that conveys the maximum amount of sample to the mass spectrometer ion source.

Unfortunately, capillary columns are limited in their sample- and solvent-carrying capacity. It is often necessary to split the sample at the injector to prevent column damage by a relatively large injection of hot solvent. To overcome the problem of sample loss due to split injections, a number of splitless injectors and injection techniques have been developed, for example, the Grob-type injector shown later in Figure 64. Operation with capillary columns requires more skill from the operator than packed columns. Capillary columns are also fairly delicate and require careful handling, although recent developments with fused quartz have led to remarkably robust columns. Figure 62 shows the extent to which there can be bent, although gentler treatment is recommended.

Retention times with capillaries tend to be long, leading to long analysis times; a 1-h temperature-programmed run is not uncommon. As a result, they are not viable in terms of instrument time unless the mixture for analysis either is very complex or involves a difficult separation. Although packed columns can handle large injections, it is undesirable to saturate the mass spectrometer with

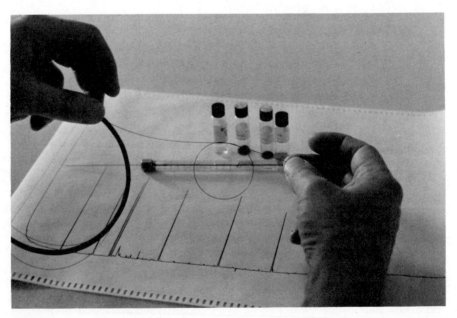

Figure 62. Fused quartz capillary columns, so robust they can be tied in knots. (Courtesy of Masspec Analytical Ltd.)

solvent, so capillary columns can sometimes offer sensitivity/detection limit improvements. The narrow-capillary gas chromatography peaks are much more intense for the same amount of material injected than those from packed columns.

Injectors

The injector arrangement of a gas chromatography system fulfills a number of purposes quite apart from the obvious one of sample introduction. The injector acts as a point of anchorage for the gas chromatography column and as the carrier gas connection to it. The assembly is usually contained in a heated block of large thermal mass. This serves to preheat the carrier gas as well as flash evaporating the injected sample.

Packed Column Injectors.   A typical arrangement for packed column work is shown in Figure 63. The carrier gas passes through the heated injector block before entering at the bottom of the injector body. It then passes up the body around the outside of the column before sweeping over the septum. This ensures that all the sample is carried into the column. Commercial columns usually have a groove, notches, or pips at the injector end. These prevent the operator from sealing the column off while pushing it up to the septum. "Home-made" columns do not always have such refinements, and care should be taken to allow a small clearance at the septum end. This is usually achieved by pulling the column back 1 or 2 mm. However, this can cause undesirable dead volume at the

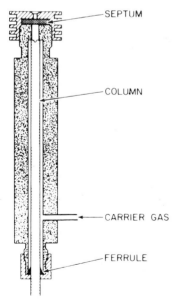

SEPTUM

COLUMN

CARRIER GAS

FERRULE

Figure 63.   Cross section of an injector assembly for packed column gas chromatography. (Copyright © 1982, Finnigan Corporation. All rights reserved)

other end of the column. It would be better to shorten the injector limb of the column slightly to achieve the desired clearance without the need to pull back the column at the detector end.

Capillary Column Injectors.   In capillary work the carrier gas flow rate is an order of magnitude lower than for packed columns and is not usually sufficient to sweep the injector body clear after injection. To overcome this problem, an additional flow of carrier gas is routed through the injector to purge the body and the septum. The flow path is controlled by glass and metal inserts within the injector body. Inevitably, this additional flow will tend to reduce the amount of sample entering the column. This feature is exploited in some designs to provide an adjustable split ratio by controlling the flow rate of the purge gas. For a split-less injection the purge lines are closed just before sample introduction, forcing most of the sample onto the column.

The injector is usually operated at constant pressure so that the column flow rate is unaltered as the purge valves are opened and closed. The Grob injector of Figure 64 illustrates a typical arrangement that can operate in both split and splitless modes. The septum sweep valve is shut just before injection and is left closed for about a minute to allow optimum sample transfer onto the column. In the split mode the sample split valve remains open at all times; a needle valve controls the split-body purge gas flow and hence the split ratio. With fixed head pressure, the flow rate through the needle valve will be constant, but the flow rate through the column will depend to some extent on column temperature. At injection it is usual to set the column 15 to 20°C below the solvent boiling point to take maximum advantage of the solvent effect. The column carrier gas flow will then depend principally on the pressure, and the split ratio will be consistent.

When the Grob injector is operated in the splitless mode, both the septum sweep and the split-body purge gas lines are closed during the injection phase. After about a minute they are opened to clear the injector assembly. Any septum bleed or contamination from the injector body assembly is then subjected to a significant split ratio and only small amounts enter the column. In this way ghost peaks and background effects are minimized as the column is temperature programmed or raised to its isothermal operating point.

On-Column Injection.   Even with splitless injection there may be some loss of sample within the injector body. Furthermore, the rapid evaporation in the heated injector followed by condensation onto the column is inefficient and prone to discriminating effects and may cause the breakdown of sensitive compounds. To overcome these problems, on-column injectors (Fig. 65) have been developed. The liquid sample is introduced through a precision guide and placed directly into the precooled column. Alignment is critical, but the procedure is no more complicated than operation of the Grob injector. The on-column injector is usually cooled rather than heated. It does not have a septum and has

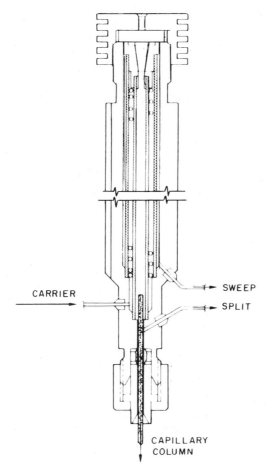

CARRIER

SWEEP

SPLIT

CAPILLARY
COLUMN

Figure 64. Grob type injector for capillary column systems. (Copyright © 1982, Finnigan Corporation. All rights reserved.)

only a small exposed area of valve seal material. Consequently, bleed levels are very low. In conventional heated injectors, careful choice of septum and minimum area designs are needed to reduce the effect of septum bleed. Figure 66 shows a typical assembly that is commercially available.

Alternative Sampling Methods. In some applications the injector assembly is modified or replaced by an alternative arrangement. For example, with batch inlets and sample concentrators/desorbers, sampling loops or valved gas flow systems are often used. Figure 67 illustrates the principles of operation of a gas-sampling loop. Carrier gas flows via the sampling valve arrangement into the gas chromatography column while the sample gas flows to waste via another loop of known volume. Operating the valve disconnects the sample gas from the sample

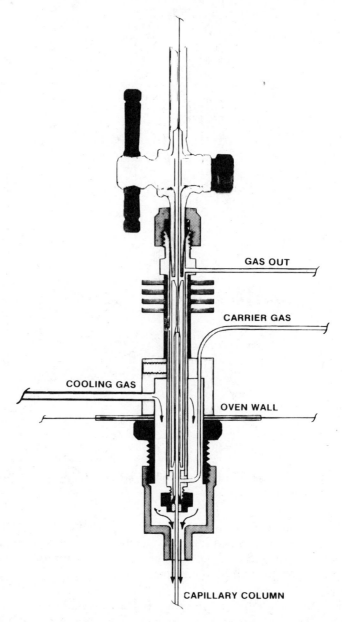

Figure 65. Cross section of an on-column injector. (Copyright © 1982, Finnigan Corporation. All rights reserved.)

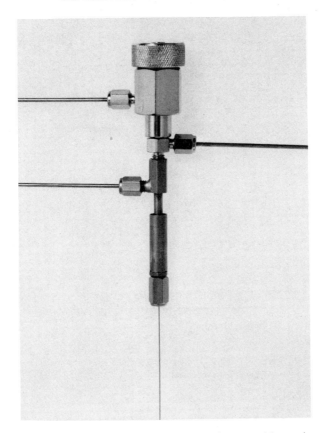

Figure 66. An on-column injector. The three connections provide carrier gas, cooling, and pneumatic operation to allow syringe entry. (Photograph courtesy of Scientific Glass Engineering.)

loop. The carrier gas flow is rerouted to carry the sample gas from the loop onto the column. This is a useful technique for sampling gas flows in reaction studies and process monitoring.

Another sampling technique involves the use of a sample concentrator. Figure 68 shows the technique applied to organics in water analysis. Helium is bubbled through the water sample and then into a polymeric trap at room temperature, carrying any volatile pollutants with it. The trap is then connected to the gas chromatograph inlet gas supply either by a valve arrangement or by physically plumbing it in. Rapid heating of the trap causes the organic materials to be desorbed and carried onto the column for analysis. By carefully controlling the gas flows and heating cycles as well as the initial chromatographic conditions, good GC/MS data can be obtained. Finnigan's OWA (organics in water analyzer; Fig. 69) is one such instrument dedicated to this type of analysis.

This technique is not limited to water sample concentrators but has been ap-

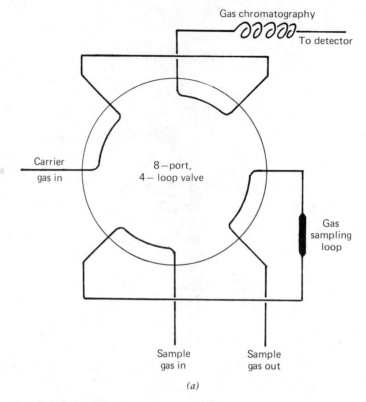

Figure 67. A gas sampling loop arrangement. (*a*) Sampling position.

plied to atmospheric sampling of airborne pollution as well (Figs. 70 and 71). The trap can be set up for sample capture almost anywhere and then sealed to prevent sample loss or unknown contamination until it is ready for connection to a GC/MS system. A similar technique using a cold loop, a loop of tubing cooled by liquid nitrogen, can also be used to trap airborne organic material.

Gas Chromatography Detectors

The detector in a GC/MS system is of course the mass spectrometer. Conventional gas chromatograph detectors such as the flame ionization detector and the electron capture detector may also be fitted. This enables the GC/MS operator to reproduce gas chromatography data sent in with samples for analysis before connecting the column to the mass spectrometer. It is important to establish similar chromatographic conditions when analyzing samples provided by another party. This is especially true when dealing with complex mixtures or particularly sensitive compounds.

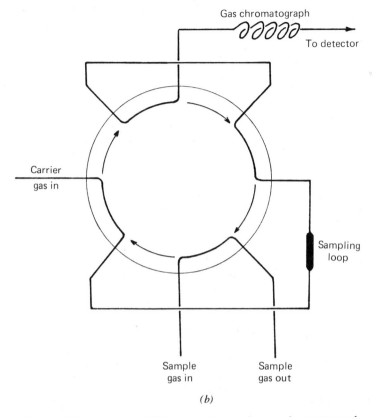

(b)

Figure 67 *(continued)* *(b)* Introduction to the gas chromatograph.

Once similar gas chromatography traces are obtained, the column can be connected to the mass spectrometer. The total ion trace should then show a pattern similar to the gas chromatography traces, but some reduction in retention times is likely. This is because most GC/MS interfaces reduce the pressure at the end of the column compared with conventional gas chromatograph detectors, which usually operate at atmospheric pressure. The lower end pressure reduces the effective length of the column and thus shortens retention times, but the overall performance should not be degraded by the GC/MS interface. Comparing the gas chromatography data with the GC/MS data is an excellent way of verifying satisfactory interface performance.

Fittings and Gas Controllers

Pipe fittings and ferrules play an important part in GC/MS interfacing as well as in the connection of gas chromatography columns to injectors. A discussion of

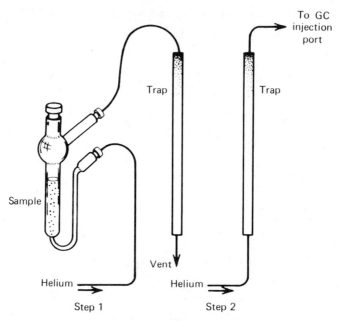

Figure 68. Sample concentration using a polymeric trap. The sample is first trapped at room temperature. The trap is then connected to the gas chromatograph and the sample desorbed by heating. (By kind permission of Finnigan-MAT Ltd.)

Figure 69. Finnigan-MAT's OWA (Organics in Water Analyser) GC/MS system. (Copyright © 1982, Finnigan Corporation. All rights reserved.)

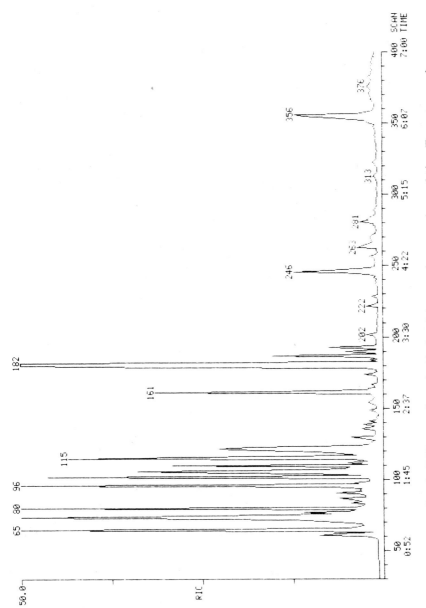

Figure 70. An RIC trace of a typical industrial atmosphere sample caught in a Tenax trap and then desorbed into a GC/MS system for analysis. (By kind permission of Finnigan-MAT Ltd.)

119

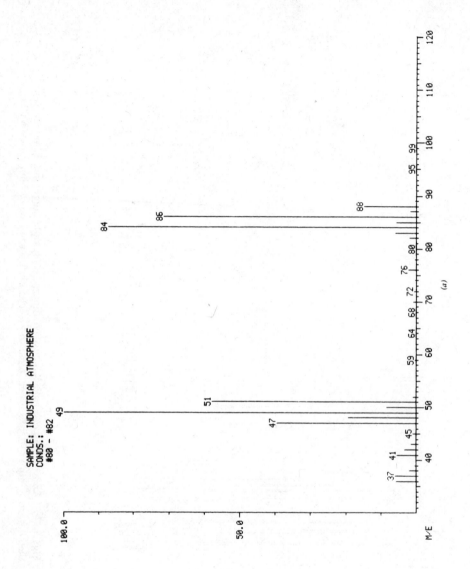

(a)

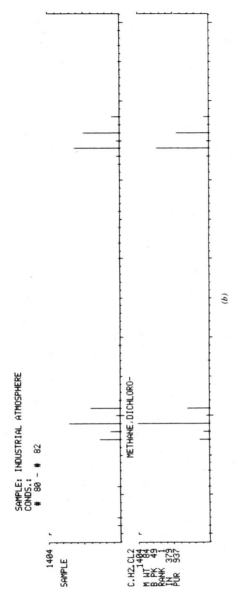

Figure 71. Analysis of data from Figure 70. (*a*) A typical spectrum. (*b*) Library search of this spectrum. (By kind permission of Finnigan-MAT Ltd.)

their characteristics has been left to the end of the chapter on interfacing so a full appreciation of their various requirments can be gained. Other pieces of gas chromatography hardware that warrant some discussion are flow and pressure controllers.

Flow Controllers.   Most packed column work is done at a constant carrier gas flow rate, typically 15 to 30 ml/min. Gas chromatography characteristics vary with flow rate, so it is advantageous to keep this parameter constant. The use of electronically controlled flowmeters is becoming commonplace. These units are generally reliable if set up correctly. In particular, it is important to keep the carrier gas supply pressure within the limits laid down by the manufacturer or erratic, even pulsing, flows may occur. It is also good practice with both mechanical and electronic controllers to fit a fine filter in the carrier gas supply line where it enters the gas chromatograph cabinet if the manufacturers have not already done so. Flow controllers are very susceptible to contamination by small particles.

   Even though the flow controller may be rated up to a relatively high pressure, it should be limited to a reasonable maximum. If a blockage occurs in the injector, column, or interface, the flow controller will allow the pressure to build up to the maximum in an effort to maintain the flow. An injector pressure meter is a useful indicator in these circumstances and can give advanced warning of flow problems.

Pressure Regulators.   The constant-flow, variable-pressure characteristics of flow controllers are unsuitable for capillary work. The changing flow requirements in and out of the injector are beyond the scope of most flow controllers. Overall, it is most practical to operate at constant pressure, even though column flow may vary with temperature and other conditions. If the system is fitted with a constant flow controller, turn the carrier gas tank pressure down and then open up the flow controller fully. The regulator on the tank is now the dominant controller, and constant pressure operation should be possible.

CHAPTER FIVE

GC/MS Interfacing

A typical carrier gas flow rate for a packed column gas chromatograph is 30 mL/min. The detector will normally be operating at atmospheric pressure. Assuming no temperature change, this is equivalent to about 10^9 mL/min of gas at mass spectrometer source pressures. The pumping speed at the ion source is unlikely to exceed 150 L/s, that is, approximately 10^7 mL/min. Clearly two orders of magnitude are too great a difference to be accommodated by reducing the carrier flow rate, and some sort of gas-removing device is required. For optimum system performance it is important that the excess carrier gas be removed but that the sample remain. In other words, a gas-sample separator is required.

Flow rates can be reduced to ease the process, 15 to 25 mL/min being the most common range for GC/MS work. The choice of gas is also important. Nitrogen, the usual gas chromatography carrier, has a much lower pumping speed than the lighter gases such as helium and hydrogen. It is not removed so easily from the vacuum chamber by the pumping system and would put greater constraints on the separator performance. Nitrogen also cools significantly when it expands into a region of lower pressure; condensation of the sample could therefore be a problem. Helium and hydrogen do not show this effect as much and are both used as carrier gases in GC/MS systems, helium being the most common because of safety considerations.

SEPARATORS

Helium, as well as having good pumping characteristics, also has a low molecular weight, which is used to advantage in a number of separator designs. For the most part the sample molecules leaving the gas chromatograph are one or two orders of magnitude heavier than helium. The diffusion rate for helium through porous materials is consequently much greater than that of the sample molecules.

The Watson–Bieman Effusion Separator

The column effluent enters the separator, as shown in Figure 72. This consists of a sintered glass tube of ultrafine porosity through which the carrier gas effuses to be pumped away by an ancillary vacuum system. The remaining carrier, greatly enriched, passes through the separator to the mass spectrometer ion source. Restrictions at either end of the sintered tube optimize flow rates and pressures for the device. A number of other materials have been successfully used in place of the sintered glass tube, silver, stainless steel, and ceramic being the most common.

The Jet Separator

The molecular weight difference between helium and the sample can also be used to advantage in jet separators, where momentum and diffusion rate are used to enrich the gas flow to the mass spectrometer. The column effluent is forced through a fine jet into an evacuated enclosure. Opposite the first jet and only a very small distance from it is mounted a second jet. Under normal gas-chromatography flow conditions the gas will pass through the jets at near super-sonic speeds. The heavy sample ions travel almost in a straight line from one jet to the other, but the lighter helium molecules, under the influence of the pressure differential, diffuse outwards to give a cone-shaped gas stream, as shown in Figure 73.

Most of the helium molecules miss the second jet and are pumped away by the ancillary vacuum system. By selecting the jet sizes and spacing, the degree of enrichment (amount of carrier gas removed) and the yield (amount of sample reaching the ion source) can be set. To achieve the required ion source pressures it is not uncommon to employ two stages of jets operating in series. Figures 74 and 75 show different versions of commercially available jet separators.

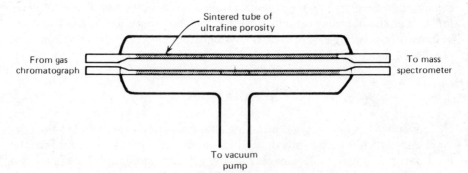

Figure 72.   Watson-Bieman effusion separator.

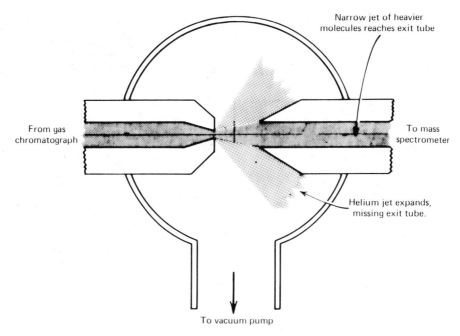

Figure 73. Jet separator—principles of operation.

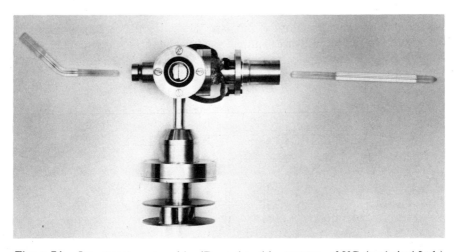

Figure 74. Jet separator assembly. (Reproduced by courtesy of VG Analytical Ltd.)

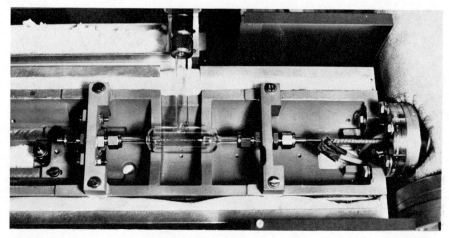

Figure 75. Jet separator in GC/MS transfer oven. Note the GLT elbow from the gas chromatograph and the heated GLT transfer line to the mass spectrometer. (Copyright © 1982, Finnigan Corporation. All rights reserved.)

The Membrane Separator

This type of separator relies on the solubility of organic materials in a silicone polymer. The inert carrier gas has a very low solubility in the membrane material, and so the amount of helium reaching the ion source is small. The membrane separator (Fig. 76) consists of a chamber across which is stretched an extremely thin silicone membrane. Membrane thickness varies with design, but 0.025 mm (0.001 in.) is common. The membrane would be easily ruptured by the pressure differential across it if it were unsupported. A fine metal mesh or sintered glass screen is used for this purpose. The outlet for the excess carrier gas can be at atmospheric pressure. Sample yield to the mass spectrometer is compound-dependent and may be as high as 60%. The rest of the sample remains in

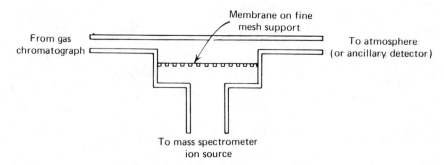

Figure 76. Membrane separator.

the gas stream and can be fed to a conventional GC detector for comparison purposes.

SOLVENT DIVERTERS AND DUMP VALVES

A large part of most chromatography runs contain little or no useful information. Even so, in these "quiet" zones a significant amount of material (carrier gas, septum, and column bleed) is reaching the mass spectrometer ion source. The amount of material goes up dramatically as the solvent sweeps off the column into the vacuum system, which causes the pressure to rise suddenly. Under certain conditions the pressure increase may be sufficient to trip out the vacuum protection circuits and may even cause gauge and ionizer filaments to burn out. Even if these problems do not occur, the material reaching the ion source is a continuous cause of contamination.

To prevent these undesirable effects it is common to fit a diverter, or dump valve as it is sometimes called, in the GC/MS interface. Figure 77 shows a typical arrangement. The valve may be operated either manually or under electronic control. The operator can now direct the effluent flowing out of the gas chromatograph either to the mass spectrometer or to waste, usually to an ancillary vacuum system but to atmosphere in some designs. Not only can the solvent peak be prevented from entering the vacuum system, but also large parts of a chromatographic run can be dumped. Allowance should be made for the fact that the diverter arrangement often places a partial vacuum at the end of the column. This may cause a reduction in retention time compared with undiverted operation. This is illustrated in Figure 78, which shows a reduction in retention time for the early gas chromatography peaks when the vacuum diverter is used. Note also that the diverter eliminates the grossly overloaded solvent signal. By using

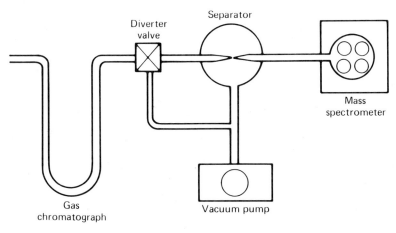

Figure 77.   GC/MS interface diverter arrangement.

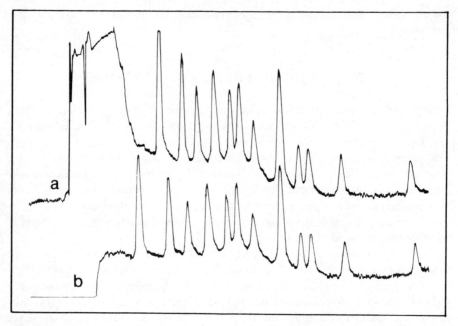

Figure 78. Effect of vacuum diverter on GC/MS response. Trace *b* shows no solvent response and shorter retention times for early peaks due to the pull of the divertor. (Copyright © 1982 Finnigan Corporation. All rights reserved.)

the diverter, columns can also be programmed up and held at a high temperature at the end of each run to clean them off without the excess bleed contaminating the ionizer. However, it should be noted that most diverters are not 100% efficient and some flow to the mass spectrometer will occur. Columns should therefore be disconnected from the GC/MS interface for major conditioning operations.

It is not normal to use a diverter system with capillary interfaces. The advantages given above are outweighed by the fact that the diverter plumbing may introduce a relatively large dead volume to the interface, which could degrade the chromatographic performance. However, some direct/split interface designs can be operated in a divert mode by setting a very high split ratio. Carrier gas continues to flow to the ion source, but the sample, including unwanted solvent, etc., is greatly reduced. The operation of capillary interfaces is discussed more fully below.

CI INTERFACES

In the CI mode the ion source is operated at a relatively high pressure. Even so, it is still undesirable to have a high flow rate of helium to the source (unless helium is being used as the reagent gas). CI sources are very tight, and the partial pressure of helium in the source can rise so much that it interferes with the action of the CI reagent gas. There are two basic solutions to this problem: either use a gas-sample separator as described above or use the reagent gas also as the carrier gas.

If a separator is used, allowance must be made for the changed pressure differential in CI operation compared with the EI mode. For instance, with jet separators it is sometimes advantageous to throttle the pumping stem so that reasonable pressure ratios are maintained across the jets. Reagent gas is then added to the enriched carrier to bring the source pressure up to the CI operating levels. It is usual to add the reagent as "makeup." That is, the plumbing arrangement adds the reagent gas in such a way that the sample and carrier gas are swept into the ion source together. This is usually achieved by a concentric flow system. The sample line runs part way through a larger bore tube connected to the ionizer. Reagent gas is added into the large tube so that it will sweep concentrically over the end of the smaller tube in the same direction as the sample and carrier gas. Adding the reagent gas to the ionizer such that it flows against or at right angles to the carrier gas flow may reduce sensitivity by blowing the sample out or to one side of the ion source, preventing full ionization. Figure 79 shows a typical combined interface design with a coaxial CI makeup-gas arrangement.

Direct connection to the ion source without a separator using the reagent gas as carrier will give maximum sample utilization and prevent carrier-reagent-sample interaction. Methane, for instance, is a good carrier gas and is an ideal general-purpose CI gas. Ammonia, surprisingly, has been used in a few cases as both carrier and reagent gas but is not recommended as it may destroy column performance. A lot of other gases, isobutane, for instance, are unsuitable carriers because their mass is too great for sufficient flow rates to be maintained at the required column temperature. With their smaller internal diameters, capillary columns suffer with this problem for all but the lightest of gases and are rarely used with the reagent as the carrier gas.

Direct connection using helium as the carrier with a reagent gas as makeup is not always an undesirable arrangement. The high partial pressure of helium in the source leads to some relatively high-energy charge-exchange spectra that are very similar to conventional EI spectra run at low electron energies. This spectrum will be superimposed on the CI spectrum. The analyst is therefore presented with both CI and EI data at the same time, often a useful combination for compound identification.

CAPILLARY COLUMN INTERFACES

Wherever possible, direct connection to the ion source is preferred for capillary columns. With the ultraflexible fused silica columns it is not uncommon for the column to be taken right into the ion source without any connections and associated dead volumes. Flow rates in the range 0.5 to 2 mL/min for directly coupled

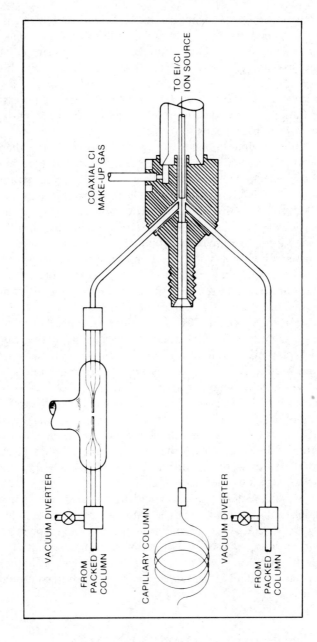

Figure 79. A typical GC/MS interface with a coaxial CI reagent gas make up line. (By kind permission of Finnigan-MAT Ltd.)

capillary columns are too high for some mass spectrometer EI sources. Source pressure can be reduced to acceptable levels either by fitting a splitter or by modifying the source pumping characteristics. Some instruments have sources with a special capillary mode. For instance, the Finnigan 4000 has a variable conductance aperture. For EI direct (capillary work), a plug is withdrawn to increase the source conductance, and flow rates up to 4 mL/min can be accommodated.

Few sources can handle the slightly higher flow rates typically used with SCOT columns. For these columns a sample-enrichment device or flow splitter must be used. There are a number of disadvantages associated with capillary splitters. Obviously, sample utilization is reduced by the split ratio, and some designs are also mass-discriminating. As the molecular weight of the eluting compounds increases, the split ratio may change. The most critical problem, though, is usually dead volumes; this can cause significant peak broadening and tailing, reducing the advantages of the SCOT column. Many people choose to use the finer bore capillary columns (WALLCOT) for these reasons.

The amount of bleed from a capillary column is usually less than with packed columns, provided a reasonable margin is allowed at the upper temperature limit, and so direct coupling without a diverter is not usually a handicap. Care should be taken to protect the mass spectrometer and its vacuum system by injecting as little solvent as possible and switching off filament and high voltage supplies until after the solvent has passed through. Columns should be disconnected from the mass spectrometer during conditioning to avoid source contamination.

One design of splitter, the open split coupling, overcomes some of the problems of other designs and can also be operated in a diverter mode. Figure 80 shows the basic arrangement. The capillary column butts up to the transfer line, which is usually made of fine-bore glass or platinum tubing. This connection within the splitter assembly is flooded with helium, and the amount of helium entering the assembly controls the split ratio. Furthermore, by introducing the helium from a small-bore tube lying alongside the transfer line, it is possible to blow away most of the eluant, so the assembly acts as a sample or solvent diverter. It should be noted, however, that the flow of helium to the source will be unchanged. Directional flow of the helium to give the split or divert modes is controlled by a solenoid valve arrangement.

TRANSFER LINES AND FITTINGS

With the advent of very flexible fused silica capillary columns, the possibility of a direct connection to the mass spectrometer source has become a reality. Even so, in many cases it is necessary to use connections, interconnecting tubes and transfer lines. In all cases the fittings and lines should be inert and of a low dead volume. All-glass systems are preferred. Glass-lined tubing (GLT) provides a less fragile alternative to ordinary glass tubing. Although robust, GLT will not stand up to cold bending without the internal glass lining being shattered. The

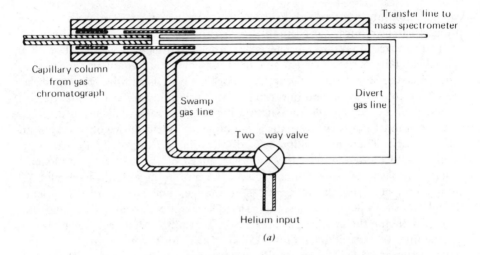

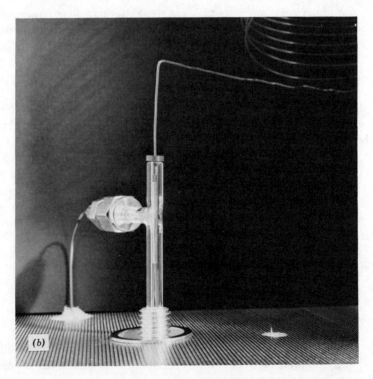

Figure 80. The open-split capillary column diverter. (*a*) Diagramatic cross section. (*b*) Closeup of a typical example. (Finnigan-MAT GmbH)

exposed metal and adhesive is then an extremely active surface. Care should be taken when adjusting fittings connected to GLT to avoid undue bending or twisting.

Dead volume is the space within tubes and fittings where the gas flow is such that there may be swirling, mixing, or a delay in sample transfer. This leads to peak tailing and loss of chromatographic resolution. Fittings and ferrules should be chosen to avoid sudden changes in pipe diameter and trapped spaces. The smallest GLT tube typically available has an internal diameter of 0.3 mm, which is too large for most capillary interfaces. Very fine-bore noble metal tubes, especially platinum and gold, have been used successfully instead. Most compounds will pass through these without degradation, but for all interfaces there may be problem compounds. In these cases derivatization of the sample may be the only solution to the problem. GLT in the range 0.5 to 1 mm in diameter is ideal for packed column transfer lines following the gas-sample separator. It is also used for the interconnection between the separator and the column in the oven of the gas chromatograph. These applications of GLT can be seen in the photograph of Figure 75.

Ferrules

The choice of ferrule for a given GC/MS fitting depends on the materials being joined and the operating temperature. Metal ferrules should be used only on joints considered to be permanent that can stand the swaging effect when the fitting is tightened; for example, gas line plumbing, but not on GLT.

Vespel Ferrules. Vespel is a DuPont polyimide material that offers a viable alternative to metal ferrules. It can operate up to 350°C and can stand repeated temperature cycling without shrinkage or leakage. Bleed levels are low, and even at high temperatures it does not flow to any great extent. Care should be exercised when tightening Vespel ferrules, especially when they are cold, as they have a tendency to shatter. The hardness of Vespel and its resistance to flow make it unsuitable in most cases as a ferrule on glass tubing. Furthermore, it will stick solidly to glass, and its removal with a knife, file, or hacksaw is a risky business, especially when the glassware is a valuable column or jet separator! If Vespel ferrules must be used on glassware, care must be taken to choose the correct size. A 1/4-in. Teflon ferrule, for instance, will flow down to a 6 mm tube as the fitting is tightened without any problems, but 1/4 in. (6.3 mm) Vespel ferrule will shatter. The correct size (6 mm in this case) must be used.

Teflon Ferrules. Teflon is the normal choice of ferrule for glass tubing. Pure Teflon flows very easily and is normally used only in room-temperature applications such as vacuum gauge connections. There are a wide range of glass- or ceramic-loaded Teflon materials that resist flow at raised temperatures, giving an upper limit of about 250°C. Above this temperature even glass- and ceramic-loaded Teflon flows too much to be usable. As the operating temperature increases the problem becomes more severe, and an alternative material, such as Vespel, must be used.

Graphite and Graphite-Loaded Teflon Ferrules. These have good flow and temperature characteristics but cannot be recommended unconditionally. There is always a tendency for graphite materials to shed very fine particles, which can easily block or obstruct transfer lines and the fine jets of separators. Graphite is also an extremely active material capable of absorbing a wide range of organic compounds. Fine particles from these ferrules are therefore very undesirable in any gas chromatography system. Graphite and graphite-loaded Teflon ferrules should be used with care to avoid these problems, and in most cases they are not worth the risks. Ceramic-filled Teflon or Vespel are usually viable alternatives.

AUXILIARY INLETS

Although a GC/MS system is dedicated to analysis of the GC effluent, most instruments will have at least one other sample inlet.

Direct Insertion Probes

Solids Probe. This is the most common form of alternative input to GC/MS systems. A solid sample is placed in a small glass vial held in the end of the probe wand (Fig. 81). The probe enters the vacuum system through a series of vacuum interlocks so that the sample cup butts up to the ion source. Heating the probe causes the sample to evaporate into the ionizer ready for analysis.

For most samples a temperature of a few hundred degrees centigrade or less is sufficient. The very low pressure of the vacuum system increases the sample volatility. Often the sample evaporates within the vacuum interlock as it is pumped down. To overcome this problem, most probes are fitted with cooling lines. For the occasional sample, adequate cooling is readily provided by a hand-held aerosol spray, which can be directed through the cooling lines. However, where large numbers of probe samples are run, plumbing the probe permanently with flexible gas or liquid cooling lines is more convenient. In either case, condensation or even ice buildup on the probe should be avoided before inserting the sample into the vacuum system. This is readily achieved by loading the sample and placing the probe in the vacuum interlock. The probe is then cooled before the interlock is pumped down. The small enclosed volume of the interlock chamber prevents significant water buildup. The vacuum system will then stabilize rapidly after insertion of the probe.

Cooling lines are also useful following sample analysis. The low system pressure prevents rapid cooling of the probe tip, and often it will be too hot to draw through the sliding Viton or Teflon vacuum interlock seals. Using the cooling lines can increase sample throughput and increase the working life of the probe

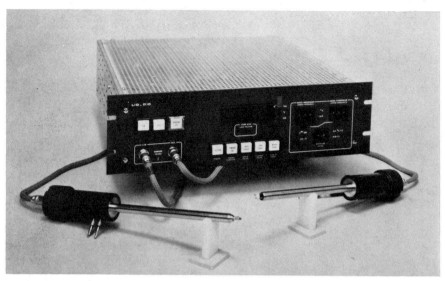

Figure 81. Auxiliary sample input units. *Left*, a direct solids inlet probe; *right*, a desorption/chemical ionization probe with their combined electronics controller. Note the provision for cooling gas/liquid flow in the solids probe. (Nermag SA)

seals. Furthermore, by combining the action of the heater control circuits and the cooling arrangement, very stable temperature control can be obtained. Fractional distillation of a sample is then possible, which enables mixture analysis. Unfortunately, volatility under vacuum conditions is often too great for adequate sample separation and the technique has limited use.

The Pyraprobe. This is a special type of direct-insertion probe capable of high temperatures and very rapid heating rates. The probe tip consists of a thin ribbon on which the sample is smeared. The pyraprobe is inserted into the mass spectrometer in the conventional way through the vacuum interlocks. The sample ribbon is then heated very rapidly to a high temperature, up to 1000°C at 20°C/ms are typical. Not only does the sample evaporate into the ion source but also it is decomposed by the heat. The resultant spectrum is of a mixture of many compounds but is nevertheless a "fingerprint" characteristic of the sample being analyzed. The technique has found application in the analysis of mold and bacteria cultures and other complex chemical mixtures or single substances that are not normally volatile.

Desorption/Chemical Ionization. Another solid-probe technique gaining rapid acceptance is *desorption chemical ionization* (DCI). The sample is flashed off a thin heated wire mounted on the end of a direct-insertion probe straight into an ionizing CI reagent gas within the ion source. Thermally labile com-

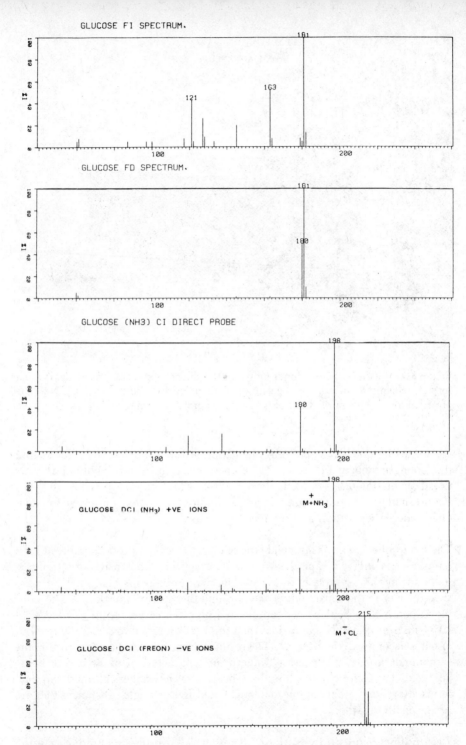

Figure 82. Mass spectra of glucose (MW 180) obtained with various soft ionization techniques. The lower spectrum shows chloride attachment achieved using DCI with Freon 113 as the reagent gas. (Reproduced by courtesy of VG Analytical Ltd.)

pounds have given strong spectra with little cracking and good molecular or quasi-molecular ions using this technique. The data shown in Figure 82 illustrate the application of DCI probes.

Batch Inlets

Direct-insertion probes are fine for solid samples but unsuitable for liquid and gas samples. They must be introduced into the system by a controlled leak. For quantitative gas analysis, precision batch inlets are available on most mass spectrometer systems; they consist of a number of calibrated chambers interconnected by a set of valves so that accurate pressure and volume ratios can be calculated before a gas sample is allowed into the mass spectrometer. To prevent contamination and background interference, the arrangement is normally placed in an oven and connected with its own pumpout lines.

Reference Inlets

More common on GC/MS systems are uncalibrated inlets for reference and reagent compounds. These usually take the form of a sample reservoir connected

Figure 83. Calibration liquid/gas reservoir and solenoid inlet control. Note also the isolation valve and pumpout control for a direct insertion probe. (Copyright © 1982, Finnigan Corporation. All rights reserved.)

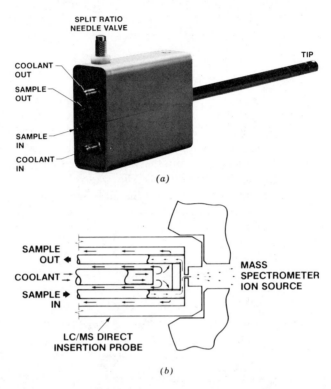

(a)

SAMPLE OUT

COOLANT

SAMPLE IN

MASS SPECTROMETER ION SOURCE

LC/MS DIRECT INSERTION PROBE

(b)

Figure 84. McLafferty type LC/MS interface. (*a*) LC/MS interface assembly. (*b*) Cross section of the LC/MS probe–ion source interface. (Hewlett-Packard Corporation)

Figure 85. Scott moving wire type LC/MS interface. (*a*) The hardware: a moving belt passing through two separately pumped vacuum interlocks before the ion source. (Copyright © 1982 Finnigan Corporation. All rights reserved.)

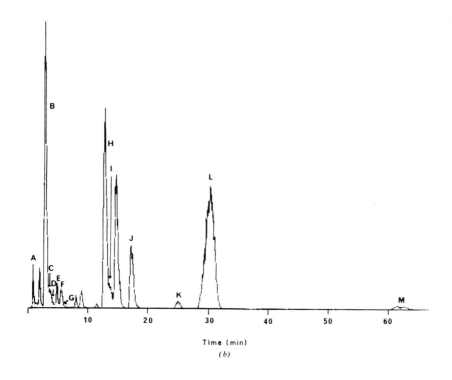

Time (min)

(b)

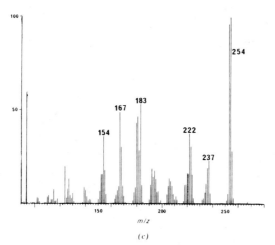

m/z

(c)

Figure 85 (*continued*) (*b*) Reconstructed total ion current trace obtained using this type of interface. The Finnigan-MAT Model 4000 mass spectrometer was operated in the EI mode. (*c*) EI spectrum obtained from peak *H* of the total ion trace, identified as elymoclavine. (from Christine Eckers, David E. Games, David N. B. Mallen, and Brian P. Swann, "Studies of ergot alkaloids using high-performance liquid chromatography–mass spectrometry and B/E linked scans," *Biomedical Mass Spectrometry*, vol. 9, no. 4, 1982. ©Heyden & Son Ltd., 1982. Reproduced by kind permission of John Wiley & Sons, Ltd.)

139

to the mass spectrometer via a shutoff valve. Gas flow is either set by a fixed leak or controlled with a needle valve. The inlet may be permanently plumbed to the mass spectrometer vacuum system (Fig. 83) or may be in the form of a probe introduced via the standard probe inlet. Some liquid samples, such as perfluoro-tributylamine, have a vapor pressure at room temperature that is high enough to carry sufficient sample into the mass spectrometer. Less volatile liquids, such as perfluoronated kerosene (PFK), must be heated to produce an adequate sample flow.

LC Inlets

Although full discussion is beyond the scope of this book, it should be mentioned that many GC/MS systems are being fitted with liquid chromatograph (LC) interfaces. There are two basic methods in current use. In the McLafferty technique the LC effluent including the solvent enters the ion source (Fig. 84). For high flow rates a splitter must be employed. The vaporized solvent acts as a CI reagent gas. The technique is confined to CI, using a limited range of solvents. Provided the pumping is sufficient, this technique is readily adapted to most instruments.

The other currently available LC/MS interface is an extension of the Scott moving-wire technique. The eluant from the LC is deposited onto a moving belt that passes through a series of vacuum stages (Fig. 85). The solvent is pumped away, leaving the sample, which is flashed off into the ion source. This type of interface can be operated in both the EI and CI modes. Solvent capacity for both techniques is limited, but with the development of micropack LC systems this is becoming less of a problem. Evaporation of polar solvents from the belt is difficult, especially with reverse-phase mixtures such as water/methanol. The problem has been partially overcome by using infrared heaters to drive off the solvent, and other techniques are under investigation.

Large quantities of polar solvents, especially water mixtures, entering the vacuum system in the McLafferty method are also a problem. Modification of the mass spectrometer vacuum system by the addition of a cryogenic pumping system (cold fingers) is usually recommended. Such a cold trap must be warmed at regular intervals so that the conventional vacuum pumps can clear the system. Consequently, continuous operation with this type of solvent for long periods is not possible.

CHAPTER SIX

Data Systems

At one time GC/MS systems were routinely operated in the manual mode. Investigation of the recorded spectra was a time-consuming business. Although interpretation of the results is a highly skilled job, much of the work involves a lot of semiskilled activities, such as cutting up LBO charts into scans, relating scans to the gas chromatography trace, record keeping, and counting and coding spectra. To speed things up and to improve the quality of the recorded material, computers have been used more and more.

Initially data were coded manually and entered into a computing system for processing. This was essentially an off-line activity. With the general level of improvements of computer systems in terms of cost, size, and speed, it became common practice to connect the GC/MS systems on-line. That is, the data were captured and stored "on the fly," and the information was then processed offline. Further improvements in computers and electronics in general opened up the possibility of computer control as well as data acquisition.

In parallel with these developments, improvements in software techniques and peripheral devices gave far greater scope for data display, manipulation, and output in real time, that is, during or shortly after data acquisition. From being an essentially manual instrument with computing facilities, the GC/MS with an integral data system has evolved into a powerful analytical tool. Indeed, some instruments have been further refined so that the data system plays a major role in parameter selection, instrument control, and data reduction. Finnigan MAT's 1020/OWA (Fig. 69) is a typical example. Without a doubt these instruments require an operator with a wide range of skills but no longer need he or she be a GC/MS expert. Many of the required operator actions are prompted by the system, and compound identification is made easier by sophisticated library search routines.

Developments have been so rapid in recent years that there are a wide variety of computer and data systems in use. Each system has its own features, and comparisons are often difficult because of the differences among systems. Our discussion must therefore be limited to the more general aspects of data systems as applied to GC/MS so that an overall understanding and appreciation of their use can be gained.

GC/MS DATA SYSTEM TERMS AND ABBREVIATIONS

Acoustic coupler	Device that links computers through the telephone system when connected to a normal telephone handset.
Acquisition	Process by which the data system takes in data from the external world.
ADC	Analog-to-digital convertor.
ALU	Arithmetic logic unit, the heart of a computer.
Batch processing	Data are gathered off-line into a storage medium. They are then loaded into the computer for processing.
Bit	Binary digit, basic number unit of a computer system. It has one of two values, on or off, high or low, true or false.
Block	A length of magnetic tape with a defined amount of data. It is usually the smallest part of a tape that can be addressed by the data system.
Boot, bootstrap program	A program that calls in a larger, more sophisticated loading program. In this way the data system is able to progressively bring its full operational capabilities to a state of readiness. From something small and insignificant the system lifts itself up by its own bootstraps. To boot a system is to effectively turn it on and make it operational.
Buffer	Area of storage in a computer and input/output devices associated with data transfer.
Bug	A fault in the software.
Byte	Originally this was a series of bits exactly half the length of a computer word. It now always means a string of 8 bits, which may not in fact be half a word of the particular computer system.
CPU	Central processing unit.
Centroid data	Data usually derived from profile data consisting of information as to a mass peak's centroid mass and its size.
Chip	Small piece of semiconductor material, usually silicon, in which electronic (often computing) circuits are made. Chips are mounted in holders and the assembly is known as an *integrated circuit*.

Command	The input instructions made by the data system operator.
Computer	A processor whose function is determined by a given set of input instructions (the program). It also has temporary storage (registers) and semipermanent storage (the memory). To be useful the computer must be able to control input/output devices.
Console	Another name for operator's terminal or VDU.
Core	Originally a memory module made up of discrete ferrite rings or cores whose magnetism indicated a binary state. Now it is more generally the memory contained within the computer accessible by purely electronic means.
Crash	Failure of all or part of a data system under fault conditions. See also *Program crash* and *Head crash*.
Cylinder	Disk systems nearly always have more than one recording head and often have more than one disk or platter. The heads are usually moved together by a single positioning mechanism. Each head is over a single track at any one time and all the heads are then said to be on the same cylinder.
DAC	Digital-to-analog convertor.
Data	Information held and manipulated by the computing (or data) system.
Data system	The combined elements of a computer, terminals, input/output devices, and storage media, including both hardware and software.
Diagnostic program	A test program that exercises part of the data system in a defined way. Used to confirm system integrity and to fault find if problems occur.
Disk	Computer memory storage medium consisting of a ferrite-coated disk that can be magnetized over its surface to store data. *Hard disks* are rigid storage disks. *Floppy disks* are thinner, more pliable disks. Their storage capacity is much less than hard disks, but handling care is less critical.
Firmware	Name given to software contained in a hardware module that is often, but not always, used to configure a piece of electronic equipment. It is usually in the form of a preprogrammed read only memory (ROM) integrated circuit.
Foreground/ background	The foreground job is the priority job of the data system. In the GC/MS system, this is usually data acquisition. Other jobs, such as data processing, may be achieved in the background, operating with a lower priority and giving way to the foreground job when necessary. In this way two or more jobs can be done by the data system on a priority time-share basis.

Hardware	The tangible parts of the computer system; cabinets, circuit boards, power supplies, and so on.
Head crash	Disk drive heads float extremely close to the disk surface. Under fault conditions the head(s) may strike the surface, causing damage to either the recording head or the disk surface. Some types of disk drive have defined areas in which the heads "land" when the disk is not spinning. A head crash in this case means a contact in the data area or an unsuccessful landing in which damage occurs. Some head crashes are classed as minor crashes in that no physical damage is caused by the contact but data may be destroyed. All crashes must be considered serious.
I/O	Input/output device.
Interface	Hardware linking the computer to the external world.
k	SI unit representing a factor of 1000.
K	A multiplication factor often used in data system work representing $2^{10} = 1024$.
Language	Computers operate under the influence of electrical signals. These signals can be translated into symbolic codes or languages that are more readily understood by the computer programmer.
Main frame	A very large computer where the capital cost of equipment and the running costs are very high. Performance is significantly better than with a minicomputer. Because of the cost, emphasis is usually on computing speed. Jobs are usually stacked or prepared in batches off-line so as to maximize the computing efficiency.
Matrix	A type of printer/plotter that produces characters by activating an array or matrix of dots.
Media	Computer storage devices. Includes tape, floppy, and hard disk.
Memory	Area of a computer where data or programs are stored.
Microprocessor	Very small processor; a computing circuit implemented in a single integrated circuit.
Mini, minicomputer	Small self-contained computer with integral power supplies. Characterized by the relatively low cost of processing data even when the capital cost of the equipment is considered. Suitable therefore for real-time on-line applications.
Modem	Modulator-demodulator. Electronic device for sending and receiving computer data via the telephone system. Electronic signals are sent directly to the telephone wires without audible communications.

Off-line	Referring to equipment not connected directly to the computer system. Off-line equipment is either operated in a local mode—that is, without the use of the computer—or data are stored in a memory storage unit for later processing.
On-line	Referring to equipment connected and ready to transmit or receive instructions or data from the computer.
Paper tape	Strip of paper punched with holes that store coded information, either programs or data.
Peripheral devices	Input/output devices connected (on the periphery) to the computer.
Processor	Electronic circuitry that manipulates electrical signals in a predefined way.
Profile data	Data points taken and stored at sufficiently frequent intervals that mass spectral peak shapes (outlines) can be established.
Program	A combination of computer instructions which when executed by the data system will cause a defined task to be carried out.
Program crash	Failure of the system software to cope with the task undertaken. It may be caused by fault conditions or by program errors (bugs). The computer either gets locked into a program loop, halts, or exits to a terminating or recovery routine, depending on design and circumstances.
RAM	Random access memory, a memory that can be read and written to.
Real time	Processing that occurs as data are fed to the computer system.
ROM	Read-only memory, a memory that can be read but not written to.
Sampling interval	Time frame in which a data point and its value are established.
Sector	A section of a cylinder on a disk system, but more often a section of a single track. In the latter case a sector is the smallest part of a disk storage unit that can be addressed by the data system.
Software	The coded instructions that control the operation of the computer system.
Tape (magnetic)	Computer memory storage medium consisting of a ferrite-coated plastic tape. Data are written onto narrow bands or tracks of the tape by magnetizing the ferrite material.
TTY	Teletype or, more generally, a keyboard printer, input/output terminal.
Terminal	Keyboard plus printer or screen allowing operator to communicate with the computer.

Track	A ring of the surface area on a disk or a narrow band on a tape where data can be recorded.
Volatile memory	Memory whose contents are lost when power to the system is turned off.
VDU	Visual display unit.
Word	Series of bits of a defined length. 8-, 12-, 16-, and 32-bit word lengths are standard.
Write protection	A switch, either real (hardware) or within a program (software) allowing a storage device to read but not write onto the storage medium.

BASIC DESCRIPTION OF GC/MS DATA SYSTEMS

A typical GC/MS data system can be considered a combination of several elements that relate to the following functions:

Instrument control
Data acquisition, storage, manipulation, and display
Hard copy output
Archival storage
Operator-computer communication

The units that fulfill these functions are the data system hardware. The hardware can often have several, even many, modes of operation. The overall control resides in the intangible parts of the system, in the software.

Instrument Control

Instrument control can mean little more than issuing a start command to a manual system. Data are then acquired as the system scans under its own internal control. The start signal may be in the form of an electronic pulse or the closure of a relay contact. In some arrangements control lies entirely with the manual system, which scans at the selected rate, giving out a signal to indicate the start and/or end of scan. These signals serve to turn the data acquisition system on and off. This type of interaction is typical of early GC/MS–data system combinations and has the advantage that the mass spectrometer control electronics are completely linear. Digital control will give discrete steps from one set value to another, although the steps may be very small. This can be considered as a linear signal with superimposed digital noise. In systems operating at high mass spectral resolution this is undesirable and will degrade operational performance. Data capture "on the fly" with the beginning and end of scan information is still a valid technique and is used in modern high-resolution systems.

For medium- and low-resolution instruments, since the step sizes are small and the instrument response factor relatively damped, direct digital control of the scan function is acceptable and in fact desirable. A typical arrangement is shown in Figure 86. Computer-generated digital signals are converted into analog levels by a digital-to-analog convertor (DAC) and are fed to the mass spectrometer electronics through a buffer circuit. The buffer may give signal-smoothing and voltage-level changes as necessary. The output analog signal can then be used to control the magnet current, magnetic field, and accelerating voltage or ESA voltage in sector instruments or the rf and dc voltages in quadupole instruments. The digital signal is generated by the computer and need not follow a linear pattern; exponential, inverse square law, or stepped (MID) functions are possible.

The operating program may also contain information relevant to instrument calibration and nonlinearities so that outputs may be given the necessary compensation for different scan speeds, resolution, and mass ranges. The digital signal is relatively immune from electrical noise and may be sent down a cable, which can be fairly long. The analog portion, on the other hand, is very susceptible to noise and pickup. Hence, it is usual to mount the DAC and analog driver/buffer close to the mass spectrometer, connected to one of the spectrometer power supplies to ensure common signal reference levels.

Not all data systems scan the mass spectrometer in small steps across the mass range. A large amount of GC/MS work is done at nominal mass. That is, the ions are assigned mass numbers rather than actual weights. If the system could jump between mass peak tops, only significant signals would be acquired and no time would be wasted in the valleys between them. However, only carbon has an integer atomic weight. All other elements show a mass defect so that compounds will not give spectral peaks at integer masses. Incorrect intensities would be recorded if the system looked only at whole mass numbers. To overcome this problem, most manufacturers using this type of scanning specify 0.1-u steps (still relatively large mass intervals in terms of available DAC resolutions) and log in the most intense sample every mass unit as the signal for the given nominal mass window. Various algorithms and operator-specified parameters are used to cope with very large mass defects to prevent incorrect assignment. Even so, care should be exercised when interpreting a completely unknown spectrum. Mass assignment errors of 1 u due to large mass defects are not uncommon.

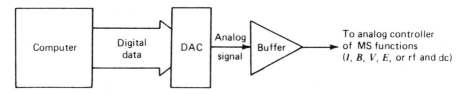

Figure 86. Digital control of mass spectrometer mass scan.

Apart from setting the mass spectrometer scan, it is becoming increasingly common for the data system to control other instrument parameters. In some instruments, such as the Finnigan OWA (Fig. 87; see also Fig. 69), all parameter selections must be made by the operator at the computer terminal. The data system requests the necessary values or selects from parameters previously stored and outputs the information to the hardware control electronics. Several sets of standard operating conditions may be stored and recalled at will. Consequently, setup times for different modes of operation can be significantly reduced. System software can also provide automatic tuning routines as well as diagnostic procedures to check and calibrate the instrument.

Data Acquisition

A variety of data acquisition techniques have been applied to mass spectrometer systems. One of the most common in use on GC/MS instruments is the sample-and-hold analog-to-digital (ADC) technique as shown in Figure 88. The GC/MS multiplier output is fed to a high input impedance preamplifier as in a normal manual system. The preamplifier is usually mounted as close to the multiplier as possible to minimize noise and stray pickup. The analog signal is then passed through a buffer amplifier to an analog integrator. The integrator is equivalent to a low-pass filter and reduces the effect of high-frequency signal fluctuations caused by noise and ion statistics. At the end of the integration time, which is normally selected by the data system, the acquired signal is transferred (sampled) to a circuit that will store (hold) the analog level. The sample-and-hold circuit will hold the signal constant so that a reliable conversion to a digital signal can be made by the analog-to-digital convertor.

During the conversion time the integrator is zeroed and then opened for the next sample. To get an accurate profile of the acquired signal, the sampling interval must be very small. Sufficient time must be allowed between samples for the ADC to output its data and discharge the integrator. Optimum signal acquisition occurs when the sampling interval and the conversion times are equal. It should be noted that in this case only half of the available signal is captured and the centroid of the stored data may be shifted. The shift may be as much as twice the sampling interval. Figure 89 has been drawn to exaggerate the effect. The sampling time is relatively long and the centroid shift correspondingly large. This effect is not normally serious at low resolution, but it may have great significance if high-resolution, accurate mass work is required. Very high speed ADCs, and therefore short sampling intervals, are used to minimize this problem. With a typical sampling interval of 50 μs, the uncertainty of mass peak centroid is no more than 100 μs. For a mass scan rate of 100 u/s, this is equivalent to 0.01 u and is normally quite acceptable for low-resolution work.

Factors other than the sampling period are often more significant in determining the accuracy of mass measurement. For example, ion statistics on low-level signals make determination of centroid mass very difficult. The fact that half the ions are lost due to the conversion delay also adds to the difficulty. To

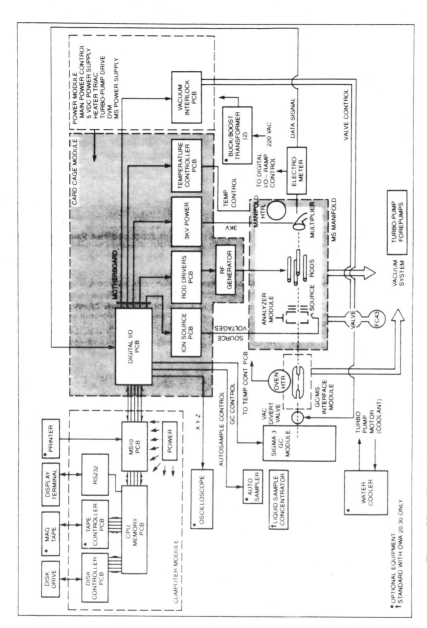

Figure 87. Block diagram of the Finnigan-MAT 1020/OWA system, showing computer control of parameter settings and hardware configuration. (Copyright © 1982 Finnigan Corporation. All rights reserved.)

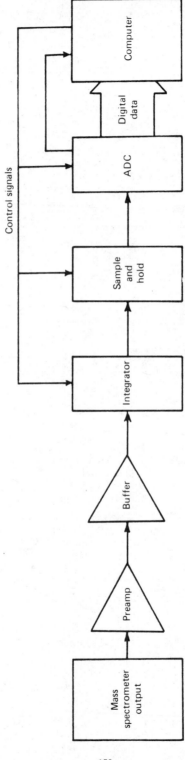

Figure 88. Data acquisition using the sample-and-hold technique.

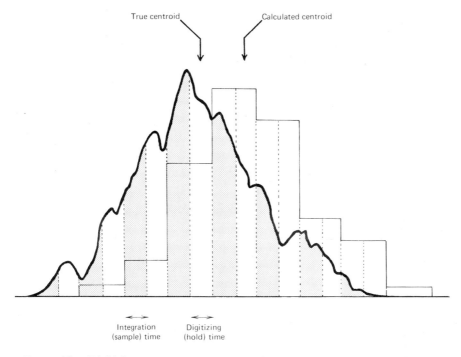

True centroid   Calculated centroid

Integration   Digitizing
(sample) time   (hold) time

Figure 89.   Digitizing a mass spectral peak using the sample-and-hold technique.

overcome these problems the Incos* data system employs two integrator–sample-hold circuits in parallel as shown in Figure 90. While one is acquiring data from the present sampling window, the other is outputting its data from the last sample window through the ADC to the computer interface. High-speed semiconductor switches are used to switch from one integrator to the other. The data are then combined so that contiguous samples are made across the mass peak. Centroid uncertainty is then reduced to only one sampling interval. Figure 91 shows the improvement in centroid calculation for the same data as in figure 89 when the dual-integration technique is used. The sampling rate in both these examples is very low to exaggerate the delay effect of the integration technique. In practice, higher sampling rates (40 kHz on the Incos system) are used, and mass accuracies of only a few millimass units are routinely possible from an instrument scanned at relatively high speed.

Also important is the ability to detect very low-level signals. When mass accuracy is not critical, integrated samples may be added to give larger sample intervals with correspondingly larger data values. The advantage of keeping the high initial sample rate is that threshold and minimum area criteria can be applied. This digital filtering is applied to the data at the interface circuit (Fig. 92) before they are stored or manipulated in the computer memory. The quality of data is

*Trade mark Finnigan-MAT Corporation.

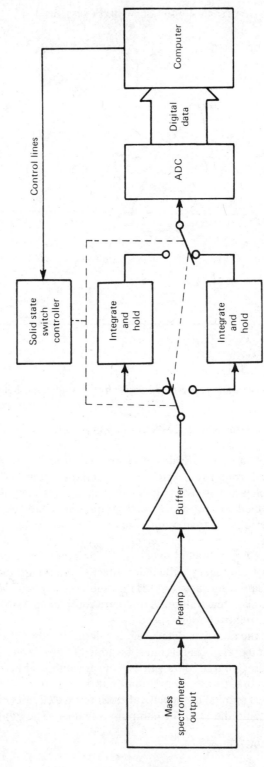

Figure 90. Double integrator, Incos data acquisition system.

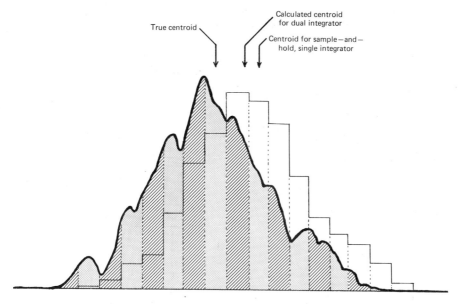

Figure 91.   Digitizing a mass spectral peak using a double-integrator system.

improved and the storage requirements are reduced by this technique. The signal is digitized in a unit mounted as close as possible to the GC/MS system before it is transmitted to the main part of the data system. This gives the greatest possible noise immunity to the signal and allows the main data system console to be sited some distance from the mass spectrometer. The hardware to implement the digital filtering and to control the remote interface is situated on a circuit board that plugs into the computer chassis.

Data Storage

A typical GC/MS instrument may scan across 300 u in 3 s, that is, 100 u/s. At first glance this does not appear to be a particularly high data rate, but assuming the data system has to calculate the mass for each ion with some degree of accuracy (rather than stepping to predetermined points), then upwards of 2000 pieces of data will be recorded each second. The computer must have sufficient working space to manipulate the data so that centroid mass and intensity values may be calculated. The data must be shifted around extremely quickly if the computer is to be efficient. The work space store or memory must be accessible on demand and capable of transferring data at very high rates. To cope with this problem, the computer has an integral memory unit, which is often referred to as the core. Ferrite cores are sometimes used as storage devices, hence the name, but more common now are the higher capacity semiconductor memories. The use of the name *core* to refer to any high-speed integral computer memory has persisted.

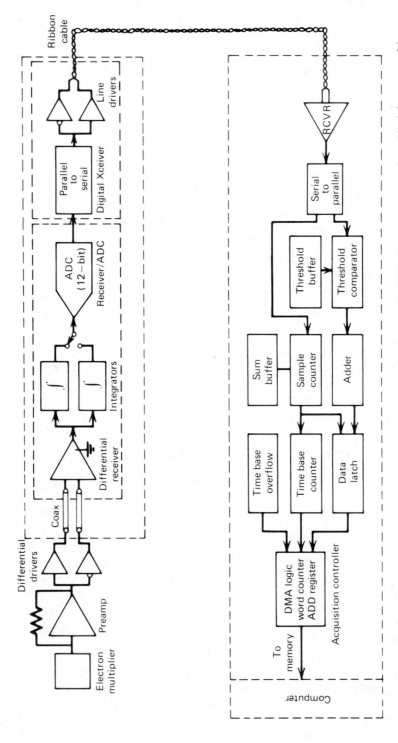

Figure 92. Incos type acquisition and digital filtering system. (Copyright © 1982, Finnigan corporation.)

154

Meanwhile, our GC/MS data system has acquired another scan of data. In fact, the data must be captured and stored as a priority over all other jobs. Early data systems did this and little else. Data processing, apart from a basic display or interscan report, was restricted to after the run had been acquired. Improvements in operating systems, in terms of either single-computer speed and efficiency or master/slave processing, has resulted in data systems with foreground/ background capabilities. The computer acquires data in a foreground or priority mode and allows processing and display of the current data (or other data) in a background, second-priority mode. In this way data can be viewed and manipulated in real time, immediately after acquisition.

Meanwhile, in our example even more data have arrived in core. In fact, it is not at all unusual for a GC/MS run to be 1000 scans long. If, after the centroid and intensity are calculated for each mass, the rest of the data is discarded, then we are faced in our example with storing 6×10^5 computer words for a 1000-scan run. Even if there is no signal over part of the mass range the data system must store a computer word of value zero to indicate this. If a word is 16 bits long, this would appear to be a great waste of space, but it is necessary if a large dynamic range of signal intensities is to be recorded. In computer terms, we require about 500K of storage, where K is 1024 or 2^{10}. This is 50 to 100 times the space normally available for GC/MS data in a typical system.

In practice the data need not be in core unless they are to be used immediately in calculations or manipulations, but they must be accessible reasonably quickly. Disk-based storage provides the necessary extension to the computer memory. After initial processing, data are written into a data file on a magnetic disk. Figure 93 shows a typical hard or rigid disk arrangement, while Figure 94 shows a floppy disk system. There is a wide variation in the storage capacities and access times of the different versions of disk storage, but the operating principles are essentially the same. Data are written onto discrete areas or tracks of the ferrite-coated disk by a recording head that floats extremely close to (but does not touch) the surface. The data are stored by magnetizing the ferrite coating; once written, they are permanent in the sense that they remain even when the power to the disk drive is switched off. Data can be erased and overwritten electronically or accidentally under the influence of stray magnetic and electric fields, and, of course, in normal circumstances they can be read back into the computer very quickly.

Disks are a relatively expensive recording medium, and where large amounts of data are to be stored it is often preferable to write the information onto magnetic tape. Magnetic tape is much cheaper than disks for a given amount of storage, and the tape is far less susceptible to damage by handling and by contamination. Reading and writing to tape is a relatively slow process, and the transport unit for large open-reel tapes is a fairly expensive piece of hardware. As a consequence there is a tendency to use magnetic tape units only where long-term storage of large amounts of data is required. Although some data systems can interact directly with data files stored on tape, this is a fairly slow process. It is more usual to read the relevant file back onto disk before further processing or

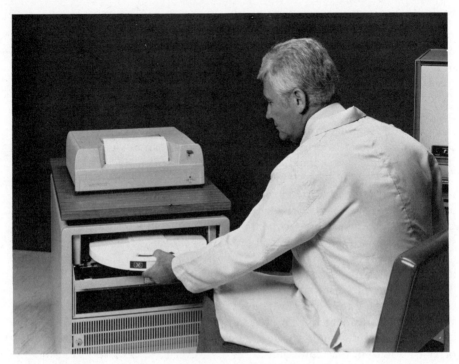

Figure 93. Hard disk data storage unit. (Hewlett-Packard Corporation)

Figure 94. Floppy disk data storage unit. (Finnigan-MAT, GmbH)

examination is attempted. the tape unit is not a replacement for a disk drive but an extension giving a relatively "slow" memory.

Floppy disks, which are less critical in their handling requirements, and tape cartridges such as the unit in Figure 95 are also used for archival storage where the amount of data to be stored is more modest. Problems in terms of software often arise where the storage capacity of the archival storage unit is less than the available space on the hard disk system. Hence some of the storage units suggested may not be supported on a given system. Data compression, enhancement techniques, and ruthless selection are the only viable alternatives if a bulk storage medium is not available.

Information can also be stored on punched paper tape or cards. These are extremely bulky and are unsuitable for large amounts of data storage, but they are used extensively to store boot-up programs, which contain a set of basic instructions for the computer so that a state of operational readiness can be easily achieved soon after the system is powered up. The operator loads the paper tape into the reader, and a simple program is then transferred into core. This in turn loads the more complex operating software, usually from the disk unit. On modern systems, boot-up or starting programs are usually stored within the computer as firmware. The program is permanently stored in specially prepared integrated circuits called ROMs (read-only memories). Boot-up procedures are then implemented automatically on power-up or by operator selection via the computer control panel.

Figure 95. Tape cartridge for data storage.

Data Manipulation and Display

Having acquired our GC/MS data it will be necessary in most cases to examine it. All data systems provide the operator with some form of display terminal or monitor on which the spectra and chromatograms can be viewed. A keyboard in the display unit, may be included or it may be provided separately. Some form of printer and/or plotter is essential, so that permanent records can be made.

Display Terminal. This may be an oscilloscope driven from the computer through digital-to-analog convertors to give the X, Y, and Z (brightness) signals. Alternatively, television-type (video) displays in either black and white or color are used by some manufacturers, but probably the most common display terminals in use are those made by the Tektronix company (Fig. 96). These computer graphics terminals work on a similar principle to storage oscilloscopes except that the display signals are formatted either as alphanumeric characters or as a graphical plot within the terminal itself from the computer-generated code. Computer-terminal communications consist of a series of digital signals, no additional digital-to-analog convertors being required.

The Tektronix terminal and many others include an integral keyboard. Where the display unit is in the form of a monitor, a separate keyboard input device is usually provided. This may be a teletype or high-speed equivalent or in

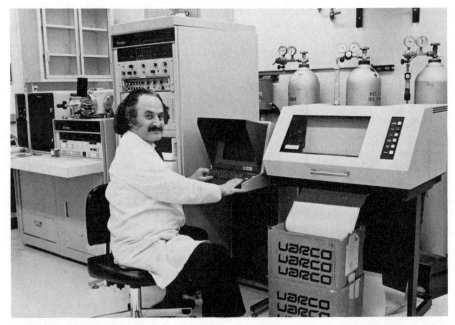

Figure 96. Computer I/O devices, a graphics terminal, a printer, and of course the operator. (Copyright © 1982, Finnigan Corporation. All rights reserved.)

the form of a custom-made keyboard assembly. Additionally, some pushbuttons and switches may be provided on separate control panels for particular functions. For example, the Finnigan 6000 series data system (Fig. 95) has about 50 pushbuttons and 8 multiposition lever switches as well as a standard teletype (not shown). Teletypes are basically computer-controlled typewriters and so include a printing arrangement. Hence, communication between the computer and the operator is either by a printed message or displayed data with or without text. The operator replies to or initiates conversation with the computer either by typing on the keyboard or by pressing one of the input pushbuttons.

Hard Copy Output.   One of the major advantages of a data system is that the data may be viewed and manipulated without creating large amounts of paper in the form of gas chromatograph traces, mass spectra, and lists. Of course, there is a requirement to create permanent records, and a number of hard copy options are usually available. Systems with teletypes as standard can readily produce typed lists. Graphical displays such as spectra and chromatograms are obtained using X-Y recorder or digital plotters. On more sophisticated systems, computer-controlled matrix or daisywheel printers are usually used. These produce high-quality typed characters at very high speed, and most units also have a graphics mode in which a dot pattern generated by the computer is used to build up a picture line by line. Figure 97 illustrates the amount of detail that can be displayed and produced on hard copy using modern computer terminals and printers.

Some display terminals have associated photocopier units that give an output that is a direct copy of the displayed information. These copiers operate off-line and require no additional computer time to create the hard copy. The copy detail is limited by the display resolution of the terminal and is not usually as fine as that obtained with line printers. The output data for the line printer in the graphical mode must be formatted differently from the display terminal data. This often takes a significant amount of computer time and slows down other processing jobs. Copiers and printers are not instant devices, and hardcopying a lot of data is a slow process. Once again, software design can minimize the impact of this. Most systems have facilities to stack data or instructions so that the actual hardcopying can be done at a more convenient time—overnight, for instance.

The nature of the displays and the dialogue between the computer and the operator will vary from system to system. It is entirely dependent on the system's operating characteristics or software.

Software

No discussion of a data system can start, let alone be complete, without consideration of its operating software. Although the basic principles are common to all data systems, the methods of implementation are many and varied. The software of a given system represents many hours of work and is therefore often as costly

PAH STANDARDS (2104)

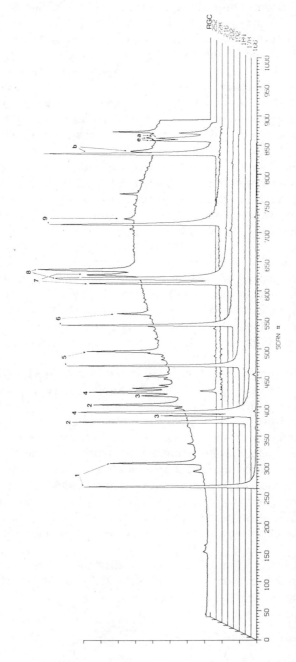

1	FLUORENE	7	1, 2 BENZOFLUORENE
2	DIBENZOTHIOPHENE	8	2, 3 BENZOFLUORENE
3	PHENANTHRENE	9	CHRYSENE
4	ANTHRACENE	B	BENZOFLUORANTHENES
5	METHYLPHENANTHRENE	E	BENZO(E)PYRENE
6	FLUORANTHENE	A	BENZO(A)PYRENE

Figure 97. Computor-generated mass chromatogram map prepared by the operator from standard data using the inherent GC/MS data system programming facilities. (Masspec Analytical Ltd.)

160

to produce as the data system hardware. Just as having the most powerful car does not guarantee success in the Grand Prix, the latest and most powerful computer hardware may not perform as well as older equipment with better software. However, before we can define good or bad software, we must first consider the design aims of the data system in question.

In terms of software, integrated GC/MS data systems fall into two categories. First there are those systems in which the software is limited at least from the operator's viewpoint. Actual system performance may be very sophisticated, but the options open to the operator in terms of instrument control and display format as well as data processing are few and are often predetermined by the manufacturer. This is not a weakness in the design of the system but a deliberate attempt to provide the power of a data system without adding to complexity of the instrumentation. In fact, by building many "if this, then that" features as well as automatic routines into the software, the overall instrument is greatly simplified even compared with manually operated instruments.

At the other end of the scale is the second main type of data system found with GC/MS instruments. This is the "all singing, all dancing" system, which allows the operator control and selection of anything and everything. It is not as demanding in terms of operator participation as one would at first imagine. In most systems, sensible default values or reasonable alternatives are provided as prompts to the operator during any dialogue. However, the operator does have

```
FINNIGAN AUTOMATED GC/MS                    MODEL  1020

    ** INSTRUMENT CONTROL COMMANDS **
CONTROL MASS SPECTROMETER .........................................  MS
CONTROL GAS CHROMATOGRAPH .........................................  GC
TUNE .............................................................  TU
MANUAL TUNE ......................................................  MT
CALIBRATE ........................................................  CA
SCAN .............................................................  SC
MULTIPLE INTERVAL SCAN ...........................................  MI

    ** DATA ACQUISITION COMMANDS **
START ACQUISITION OF N SCANS OF DATA, DISPLAY RIC ...  AC FILE #N
VIEW RIC OF CURRENT ACQU. (TO SCAN N, VERT. SCALE M) ..  V FILE #N (M
STOP ACQUISITION .................................................  ST

    ** DATA DISPLAY COMMANDS **
       (APPEND H FOR HARDCOPY--LH, ELH, ETC.)
LIST NOMINAL MASSES J TO K OF SCAN N ................   L FILE #N (J, K
LIST ACCURATE MASSES J TO K OF SCAN N ...............  LM FILE #N (J, K
LIST MASSES, WITH AVERAGING .........................  LA FILE
LIST ENHANCED MASSES J TO K .........................  EL FILE #N (J, K
SPECTRUM OF SCAN N, MASSES J TO K ...................   S FILE #N (J, K
SPECTRUM, WITH AVERAGING ............................  SA FILE
ENHANCED SPECTRUM OF SCAN N, MASSES J TO K ..........  ES FILE #N (J, K
SPECTRA OF SCAN N, ORIGINAL AND ENHANCED ............  DS FILE #N
RIC OF SCANS M TO N .................................   R FILE (M, N
CHROMATOGRAM OF MASS J, SCANS M TO N ................   C FILE (J, M, N
CHROM. OF MASS J AND RIC OF SCANS M TO N ............  CR FILE (J, M, N
RUN CONDITIONS .....................................  CO FILE

FOR COMMANDS ASSOCIATED WITH LIBRARIES TYPE "LIBR?",
QUANTITATION AND SYSTEM UTILITIES TYPE "QUAN?"
```

Figure 98. Command list for the Finnigan-MAT 1020 instrument. Most routines and procedures are preset, allowing restricted operator selection and control. (Copyright © 1982, Finnigan Corporation. All rights reserved.)

```
MSDS EXECUTIVE COMMAND FORMAT:
PROGRAM [DATA FILE][#SCAN NUMBER][,CALIBRATION FILE][/MODIFIERS]

MSDS EXECUTIVE PROGRAM NAMES:

            DISPLAY PROGRAMS: (/C,J,K,O)
LIST, SPEC, LIBR, COMP         ;SINGLE SCAN
HLIS, HSPE, HCOM               ;ACCURATE MASS SINGLE SCAN
DLIS, DSPE                     ;DUAL SCAN
CHRO, RIC, MAP, HMAP, SEAR,DCHR ;MULTI-SCAN
QUAN, RESP, EDRL, EDQL, PRIN   ;QUANTITATIVE REPORTS

            DATA PROCESSING PROGRAMS:
CALI, HCAL, RCAL, ALIG, MAGN   ;CALIBRATION (/D,E,I,P)
MASS, HMAS, DMAS, ADD, ENMA    ;SINGLE SCAN PROCESSING (/N,R,S,A,W,T)
ENHA                           ;MULTI-SCAN PROCESSING

            INSTRUMENT CONTROL AND DATA ACQUISITION:
SYST, SCAN, MID, ACQU, DACQ    ;SCAN AND ACQUISITION

            INSTRUMENT DIAGNOSTICS: (/P)
PROF, INST, RESO, GAIN         ;PROFILE DATA
FIT,  DFIT                     ;CALIBRATION
STAB, DSTA                     ;MULTI-SCAN

            UTILITY PROGRAMS:
STAT, FILE, PARA                                ;FILE MANAGEMENT
EDRT, EDCT, EDSL, EDQL, EDLB, EDLL, EDNL, EDRL  ;SPECIAL FILE EDITORS
SAVE, SAVC, SAVP, SAVF         ;WRITE FILES (.MI,.CY,.PB,.FO) (/B,X)
PRIN, TIME, ERRO, TRAC, CONV, TRAN, COPY, EPA, IDOS, RUN    ;OTHER
LOOP, QUIT, RETU, IF, $, $1, ..., $9            ;PROCEDURE UTILITES
TYPE, PAUS, WAIT, KEEP, NULL, ERAS, BEEP, FEED,  =, ?(/HJ)
SETN, SETL, SETS, SETQ, SET1,...,19, GETN, GETL, GETS
DRAW, LIFE, DOT,  SONG, PIANO                   ;FUN AND GAMES
```

Figure 99.   Command list for Finnigan-MAT's Incos data system. Each command in fact calls a detailed program with a full command structure, which allows detailed operator intervention. (Copyright © 1982, Finnigan Corporation. All rights reserved.)

the option to select alternative parameters and so retain detailed control of the system.

These two extremes can be illustrated by considering the command lists shown in Figures 98 and 99. Both were obtained by typing **?** on the instrument keyboard. Many of the commands in Figure 99 have a long list of subcommands associated with them. For instance, the command list of Figure 100 was obtained by typing **CHRO?** on the instrument terminal. Of course, there are many variations on these themes, and many systems lie between the two extremes outlined. In fact, most manufacturers offer a range of GC/MS data systems in terms of both hardware and software. It is often possible to upgrade from the basic system to provide more extensive features.

The main software elements of all GC/MS data systems will include most of the functions listed below.

Mass spectrometer tuning and calibration
Control of scanning

Data acquisition

Display presentation

Library searching

Diagnostic, self-test, and troubleshooting routines

Operator programming, or at least sequencing, of standard commands

Automatic processing: survey searching and quantitation routines

In addition, many GC/MS data systems will offer the operator standard computer facilities such as record keeping, accounting files, and programming in Basic, Fortran, or other high-level languages.

Additional Facilities

Apart from controlling the GC/MS instrument, many data systems have other hardware options. Interfaces for gas chromatography integrators, digital voltmeters, thermometers, and other data-logging instruments are typical exam-

```
CHRO ACCEPTS THE FOLLOWING COMMANDS

D(+, -, *) J,K(,L)          ,DISPLAY SCAN J TO SCAN K, (L SCANS PER PAGE)
H(^, ', ") J,K,L   (W J,K,L) ;HARDCOPY (COMMENT AND HARDCOPY) AS FOR 'D'
H! J,K,L   ,TABULAR HARDCOPY 3 GRAPHS MAX.   (CHRO/K OUTPUTS TO CHRO. 99)
(-) A,B ;(DEL) ADD MASS A; IF(B>A) A TO B, IF(B<= 5) A+/-B, ELSE A TO A+B
(-) A,B,C,D ;(DELETE) ADD MASSES EVERY C AMU FROM A TO B, D=+/- WINDOW
-          ,DELETE ALL MASSES           ,IF(D=0) D=MIN( 5,C/2)
$(XY,K)(,C)      ;ADD QUANT  MASS FROM LIBRARYXY, ENTRY K (MASS TOL=C)
/(XY,K)(,C)      ;ADD ALL MASSES FROM LIBRARY (UP TO 50 MAY BE IN TOTAL)
&(XY,K) ;SET SCAN NUMBER FROM RET. TIME IN LIBRARYXY, ENTRY K
%(XY,K) ;SET SCAN NUMBER FROM CURRENT SCAN AND REL. RET. TIME
#(XY,K) ;LABEL DISPLAY WITH NAME FROM LIBRARYXY, ENTRY K
         ,FUTURE A<(>) OR M<(>) WILL INCLUDE THIS NAME
COMPLEMENT THE FLAG CONTROLLING:
T - TOTAL DISPLAY       F1 - SAT(') MULT(_)      F - LINE,BAR,DOT FORMAT
V - INDEP. NORM         F2 - REL. RET. TIME      O - LINEAR,SQUARE ROOT

M( ,=) (J,K,L)  ,MANUAL QUAN (TO BASE, INTERCEPT)  (SCANS J TO K, GRAPH L)
M< (>)          ;WRITE (APPEND) LAST 'A' OR 'M' MEASUREMENT TO  QL FILE

R  J            ,COMP DISPLAY OF TRAILER GRAPH J, J=0 FOR RIC
B J(,K)         ;EXPAND BY J (ON GRAPH K)
U J,K           ;BASE LINE SCANS (>PEAK), MULTIPLET RESOLVER SCANS (<PEAK)
N(<,>) J,B      ;LABELS: MIN PEAK WIDTH, NOISE MULT (WRITE  SL)
A(<,>) J,B      ;AUTO QUANT: TAIL SCANS, NOISE MULT (WRITE  QL)
G(+,-,*) J,K    ;QUANT OR LABEL ONLY BETWEEN SCANS J AND K
X(<,>,*) (J)    ;USE CROSSHAIRS TO READ THE SCAN NUMBER, IF(J), SET TO J
                ,< (>) WRITE (APPEND TO) SCAN LIST, * FOR MULTIPLY BY J
K (P)    ,KEEP (GET) PARAMETERS       I        ,INITIALIZE PARAMTERS
E (Z)    ,EXIT TO MSDS                Q        ,HARDCOPY THIS PAGE
@XY      , DO METHOD CHROXY. ME
L  (S) (Y) (C) (',",*)  (J,K,L,M,N,O)    ,GO TO SINGLE SCAN DISPLAY
CR       ; 'CARRIAGE RETURN', GO TO NEXT PAGE OF DISPLAY IF POSSIBLE
```

Figure 100. Command list for the Chromatogram program of the Finnigan-MAT Incos data system. (Copyright © 1982, Finnigan Corporation. All rights reserved.)

ples. In some cases it is now possible to use the GC/MS data system as a central laboratory computing controller.

Communication options are another feature that is often available. These may be modems or acoustic coupler links for remote diagnostic and trouble-shooting or data links between associated establishments. Special interfaces with appropriate software are also available for preprocessing data prior to transmission to large data base systems, usually a centralized mainframe computer. The growth of data systems in terms of both hardware and software will continue to increase the range of facilities available to user of scientific instrumentation.

PART TWO

ROUTINE GC/MS OPERATION, TECHNIQUES, AND PROCEDURES

CHAPTER SEVEN

Gas Chromatography Methods and Techniques Relevant to GC/MS

Almost any gas chromatographic method can be transferred from a stand-alone gas chromatograph to a GC/MS instrument. The resultant TIC and RIC traces are usually quite similar but can sometimes appear to be very different. Retention times, sample-transfer efficiency, and detector sensitivity are the major factors determining the relative performance. Normal gas chromatography methods employ detectors operating at atmospheric pressure, whereas the mass spectrometer system is under vacuum. Flow rates may be different; in fact, it may not always be possible to operate with the same carrier gas. Just as the conditions must be optimized for the column-detector system in stand-alone gas chromatography, so too must the parameters and conditions be selected to give the best overall performance in GC/MS operation. As with most things in life, compromise is a necessity and trade-offs must be considered.

The use of a mass spectrometer limits the choice of carrier gas and its flow rate. Interface temperatures may set upper temperature limits for the column even though it may be capable of higher temperature programs. Even with sufficiently high interface temperatures, some columns may bleed too much to be satisfactory when coupled to a mass spectrometer. On the other hand, mass spectrometers as detectors can offer significant advantages in terms of specificity and detection limit in a particular analysis.

There are many excellent texts discussing chromatographic methods and techniques in much greater detail than is possible here. Our discussion in the main will be limited to a consideration of the compromises and trade-offs and an examination of the techniques that are unique to GC/MS systems. However, so that we do not forget that we are dealing with an instrument whose performance depends on good chromatographic practice, some basic points will be discussed in greater detail.

CHOICE OF COLUMN PHASE

It is very rare that the sample under analysis is completely unknown. In many cases the sample will demand a particular type of analytical approach. Complex mixtures with many components can only be resolved using the high resolution available with capillary columns. Packed columns are better suited to simpler mixtures, but even very simple mixtures can pose a difficult analytical problem when the components have similar retention times. Choice of a suitable column phase can usually overcome this problem, but the operator does not always have a free hand.

All columns bleed; that is, some of the phase enters the gas stream and is detected, giving a steady baseline offset. This offset is easily subtracted or backed off when the chromatogram is considered. However, the bleed constitutes a significant amount of material entering the ion source in GC/MS systems and results in a mass spectrum. The bleed spectrum may have many strong ions that can interfere with or mask small but significant mass spectral peaks from the sample. Interpretation of spectra is confused by the presence of these unwanted ions. Where computer systems are in use, background subtraction or signal-enhancement techniques are easily accomplished, but there is always a danger that small sample ions will be lost and mass ratios distorted. To minimize these interferences, the use of well-conditioned columns with lightly loaded packing materials is recommended. In some cases a different phase with a different bleed spectrum must be chosen. This is especially true in the case of quantitative methods were mass-ratio distortions must be minimized.

Another consideration is the nature of the bleed material and its ion fragments. Many of the common gas chromatography phases are silicone polymers. These break down in the ion source into a number of insulating materials, including silica. Even slight contamination of the ion volume, the ion source lens plates, or the mass analyzer by this insulating material can distort spectra and reduce sensitivity. The degree of contamination that will be produced by different phases is not easily predicted; similar materials can vary widely in their effect. When trying a new or less common phase, system performance should be monitored to assess the rate of spectrometer contamination.

THE USE OF PACKED COLUMNS

Packed columns are by far the most common type of column in use. The high flow rates sometimes used in stand-alone gas chromatography pose a problem for GC/MS systems, and flow rates of 15 to 25 mL/min are usually preferred. Even at this rate a sample-enrichment or separator device is usually needed to remove the excess carrier gas. Few systems have the pumping capacity to cope with direct connection. The nature of most separators is such that a light gas, usually helium, must be used. No separator is 100% efficient, and some loss of sample with a corresponding reduction in overall system sensitivity is inevitable. It is often worthwhile experimenting with different flow settings to get the best system sensitivity.

It is now that the column bleed spectral peaks become friends rather than enemies. Observation of the bleed spectra at some suitable high masses, say 207, 208, and 209 u, for an OV1 phase, for different flow rates with constant column temperature will reveal the optimum flow. If sensitivity continues to increase with flow, then the only solution is to operate at the maximum rate that the mass spectrometer will tolerate.

It should be noted that the column bleed will enter the mass spectrometer under the same conditions as the sample, in terms of pressure and flow through the ion source. Tuning the mass spectrometer by observing the bleed spectra will give the best system sensitivity performance in the mass range where they are visible. Calibration gases at best will be present under different flow and pressure conditions and at worst can themselves cause a lot of contamination problems. On the other hand, they usually provide a tuning reference across the whole mass range. We are stuck with the bleed and background spectra, and so we should use them to advantage for routine fine-tuning. Calibration tuning should be done only as often as seems necessary, depending on the required degree of mass accuracy and the drift of the system.

Column bleed and background spectra are very diagnostic of system performance and should be monitored routinely. The operator should become familiar with the normal background spectra for each type of column used. (Note that this advice is applicable to all types of columns, not just packed columns.)

Diverters

Packed column systems are usually fitted with diverters. The primary function of the diverter is to dump excess solvent as it leaves the column following an injection. Diverters are also useful in reducing the amount of bleed reaching the ion source. It is recommended that the diverter be used whenever the system is inactive, that is, when a column is connected but no analysis is in progress.

The divert action is often controlled by an electrically operated solenoid valve. The valve will become quite hot if left continually powered. Additional cooling of the solenoid coil by increasing the air flow around it, by increasing the coil body surface area with a heat sink, or by specifying a warm-air condition, continuously rated coil will greatly enhance solenoid coil life. These procedures also prevent degradation of the sealing seat due to excess temperature.

Diverters can also be used to good effect when analyzing mixtures with one or two highly concentrated components and some much smaller ones. A *relatively concentrated sample* can be injected and the major peaks diverted out of the system, preventing an overloaded response. Similarly, unwanted parts of the GC run can be diverted. For instance, it may be desirable to continue a temperature program for some time after the elution of components of interest to clear the column of other materials. This is, in effect, a minor conditioning operation. Major column conditioning should be done with the column disconnected at the separator/mass spectrometer end to prevent unwanted contamination; the diverter never diverts all the eluting material.

A word of caution regarding the phrase *relatively concentrated sample*. It is sometimes necessary to inject quite concentrated samples into the gas chromatograph so that very small components can be analyzed. The major components must pass through the column before reaching the diverter (or the mass spectrometer if the diverter is accidentally left closed). There is always the possibility of overloading the system, and care should be exercised.

Emphasis was also placed on *concentrated* rather than *large* samples. Samples are usually contained in a solvent, and large injections of solvent can strip off a lightly loaded phase, exposing support material. This can then cause the column to tail badly. The pumping speed of the diverter is usually too low to remove the solvent efficiently, and the mass spectrometer vacuum protect circuits are liable to trip out. In general, injections of a few microliters or less are to be preferred.

Diverters and mass spectrometer interfaces usually cause a reduction in pressure at the end of the column. As a consequence, retention times and column efficiency may be reduced compared with stand-alone gas chromatographs. There is also the possibility that there will be significant backstreaming of pump oil vapor through the divert lines. Considerable improvement can often be made by fitting foreline traps to both separator and diverter rotary pumps. Very often the same pump is used for both functions and only one trap is needed. There will be a heavy throughput of solvent, but the high flow of carrier gas is usually sufficient to keep the divert rotary pump purged. However, the foreline trap may require frequent bakeout and/or purging to maintain its performance. A lot depends on the type of solvents used and the number of samples injected.

Temperature Considerations

At one time it was accepted that the interface temperatures should be 20 to 30°C hotter than the upper temperature of the oven and that the transfer line to the mass spectrometer be hotter still. This achieved a temperature gradient between the column and ion source that prevented sample condensation. Unfortunately, high column temperatures call for very high interface temperatures and can cause sample pyrolysis. Problems with ferrules creeping and air leaks are an additional consequence. Although care must be taken not to let the interface get too cold, it is not usually necessary to have this rising temperature gradient. The pressure reduction across the separator and interface assembly is usually sufficient to prevent condensation of the sample. Separator and transfer-line temperatures equal to or even 10°C below the upper column temperature are satisfactory in most cases.

It is not a good idea to cycle the interface oven temperature, as this tends to cause leaks at all the ferrule joints. Interface temperatures are usually set to a value sufficient for most of the chromatography methods to be used. In some cases the interface may be significantly above the operating temperature of the column. This should not be forgotten as a possible cause of sample degradation when dealing with particularly labile compounds.

Septum Bleed

Just as column bleed can interfere with sample spectra, so too can bleed from the septum. The problem can be made worse when temperature-programming methods are used. Taking the temperature up and down can cause buildup of the septum bleed by condensation in the column. During the next temperature program run, the concentrated septum bleed will elute as a chromatographic peak, albeit with poor chromatography.

These baseline changes and ghost peaks are confusing and often interfere with sample spectra. It is advisable, therefore, to allow an isothermal period at the end of the run. Cooling down ready for the next sample should be done before injection, allowing just sufficient time for stabilization of conditions. When a column is to be left in an operational state for some time, say overnight, it is advisable to set it at an intermediate temperature so that condensation and buildup of interfering compounds is minimized but the column bleed level is also held down.

Buildup of contaminants from the gas tank and gas lines can also be a problem. Fitting hydrocarbon traps and scrubbers in the line just before the injector is often worthwhile. Contamination from supposedly pure gas tanks can be quite high, especially when the tank is nearly empty and its pressure is low. The problem is also exaggerated by long gas lines to the GC/MS instrument.

THE USE OF CAPILLARY COLUMNS

The resolution of packed columns is limited by the multipath effect, the different paths available to the sample through the column broaden the chromatographic peaks. Consequently, attempts to increase resolution by lengthening the column show rapidly diminishing returns. Capillary columns, on the other hand, have virtually no multipath effect, and although performance in terms of plates per meter may be less than with packed columns, long lengths can be used. As a result, very high overall chromatographic resolutions can be achieved.

Quite apart from the advantages of high separating power, capillary columns may also give a three- to fourfold increase in sensitivity compared to packed columns. Their high resolution means that the chromatograph peaks are very narrow, the sample components being concentrated into a smaller time window and the sample rate into the mass spectrometer correspondingly greater. It is important to scan the mass spectrometer fast enough or distortion of spectra may occur due to the changing concentration as the sample elutes from the column.

High sensitivity is available, and capillary columns are sometimes used even for simple mixtures to gain this advantage, but the sensitivity is won at a cost. The data rate is high and can cause a storage problem on long runs. With complex mixtures there is no easy answer. The column must be programmed slowly and the space must be made available in the data system for data capture. Where the sample is less complex, multiramp temperature programs can save considerable time and data storage. After injection the column is programmed at a fairly fast rate up to nearly the temperature range of interest. The temperature program rate is then reduced so that good separations can be achieved. A similar technique can be used to shorten the middle range of a run if the components of interest are grouped at the beginning and end of a program. An initial slow ramp is followed by a higher rate ramp, and then a slow ramp is used to acquire the higher-temperature components.

Capillary columns used on manual systems give the operator a lot of work if the spectra are selected for recording on a UV chart as they elute from the gas chromatograph. It is very easy to miss the top of a peak, which may elute in less than 3 s. One possible approach is to leave the UV paper running continuously or at least about the retention time of interest. One can then cut it into spectra, selecting the required scans and using the rest for background peak monitoring, but of course this is very expensive in recorder paper. The high cost of UV paper and the limited length that can be loaded into a recorder restrict the usefulness of manual GC/MS instruments operated in a capillary mode. Even so, the advantages offered by the capillary column make the expense and effort needed on existing manual systems worthwhile in some cases. Few people would consider it viable to purchase a new system these days with only a UV recorder, data systems being preferred.

Care must be taken if the full potential of capillary columns is to be realized. Dead volume must be minimized to prevent swirling of the sample and remixing of the components. Injection techniques are also critical to the overall performance. These factors are discussed in more detail below.

Dead Volume

If the conductance is high, then the rate of cleanout of a dead volume is also high. There will be little opportunity for components to remix as they pass through this volume. This is why it is rarely a significant problem with packed columns. The conductance and flow rates in capillary systems are much smaller, and as a consequence dead volumes must be minimized. All fittings and connections must be considered carefully and avoided wherever possible.

Transfer lines to the mass spectrometer and their connections often cause problems. Direct connection of the column through the interface assembly into the ion source is now a practical proposition with the very flexible fused silica columns. Positioning relative to the ion source is quite critical. It is possible to push the column in too far, so that sample ionization does not take place or is inefficient. Care should be taken to optimize sensitivity by adjusting the column position. It is usually possible to loosen the interface fitting just enough that the column can be moved in or out while sensitivity is monitored by the column bleed spectra.

Direct connection to the vacuum system means that it must be shut down when columns are changed. Problems can occur if the column breaks near the mass spectrometer end (this is a major worry with all-glass capillary columns). The resultant air flow into the vacuum enclosure is liable to trip all the protection systems. To overcome this problem it is common to use a short length of very fine-bore tubing, either GLT or noble metal, as an interface into the vacuum chamber. The dimensions of these interfaces are such that the vacuum system can cope with the limited air flow when the column is disconnected either deliberately or accidentally.

Hence, some connections may be necessary. To prevent peak broadening and tailing they should be made with as little dead volume as possible. "Zero dead volume" fittings are not always the best answer. These in fact may have a very small central section of small bore, which is used to locate the two halves being joined. Some proprietary zero dead volume fittings also have an internal thread so that the part used to compress the ferrule is in effect a bored-through bolt (Fig. 101a). On disassembly the ferrule is often found to be jammed inside the fitting, trapping the column. It is then necessary to break the column to disconnect it, and the fitting can be reused only when the ferrule and remaining column endpiece are drilled out. Standard fitting/nut/ferrule arrangements (Fig. 101b) usually leave part of the ferrule exposed on disassembly which can be gripped and pulled so that disconnection does not break the column.

There are a number of ways to overcome the dead volume inside the fitting. Ferrules alone do not usually fill the space. In fact, capillary column diameters vary quite a lot, and ferrules of the right internal diameter may not be available. One can drill out blank ferrules or squeeze them down by tightening the fitting, but neither of these solutions is very satisfactory. Applying a short length of shrink Teflon tubing over the end of the column is a good way of increasing column diameter to fill out the fitting. The Teflon also acts as a cushion between the ferrule and the glass. Standard ferrules can then be used with confidence. The dead volume due to the short locating boss inside the fitting is best removed by drilling the fitting right through at the standard diameter, $1/16$ in. being the most usual for capillary work.

The bored-through arrangement is also useful for joining two pieces of capillary column. For instance, it is reasonable to use a short length of very flexible fused silica capillary made from a general-purpose column as a GC/MS interface. This can be connected to glass columns of different phases as required by the analytical method and the arrangement can provide a satisfactory chromatographic surface right to the ion source. The two columns are joined by butting them together. They are held in place by a piece of shrink Teflon, which is

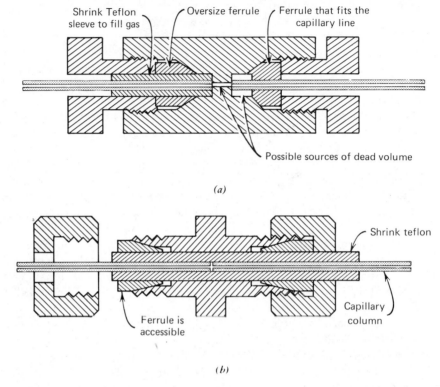

Shrink Teflon sleeve to fill gas

Oversize ferrule

Ferrule that fits the capillary line

Possible sources of dead volume

(a)

Shrink teflon

Ferrule is accessible

Capillary column

(b)

Figure 101. Capillary connections. (a) Internal ferrule. (b) External ferrule.

shrunk down onto them by gentle heating with a small flame. The join can then be held fast by a standard bored-through fitting with standard ferrules. It is important that the column ends be cut square so that they butt together cleanly, leaving no dead volume (Fig. 102). This is also important where columns are being joined to transfer lines or fitted into injectors. The shrink Teflon should also be flush with the ends of the column and two layers used if necessary to bring out the diameter to fill the fitting.

Capillary Injection Techniques

At one time it was usual to use a split technique when injecting into capillary columns. Column technology was such that only small amounts of solvent could be tolerated if the danger of phase stripping was to be avoided. A 1-μL injection with a split ratio of 10 or 20 to 1 was typical. The overall sensitivity was obviously reduced by this ratio. Furthermore, splitters in capillary injectors do not always have a uniform ratio for all compounds; molecular weight effects are not uncommon.

Modern techniques of manufacture have produced capillary columns with

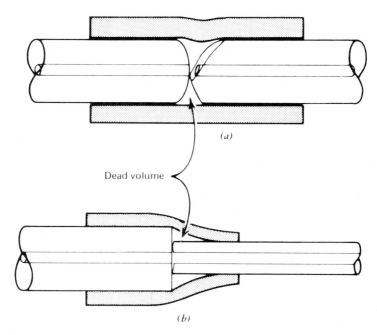

Figure 102. Sources of dead volume in butt-joined capillary columns. (*a*) Uneven butt joins. (*b*) Different outside diameters.

well-bonded liquid phases, and solvent stripping is no longer a great problem. Injections of several microliters are now possible. Split injections are no longer necessary, although it is desirable to have a split facility for quite a different reason. As will be discussed, it is normal practice with some injector designs to start the gas chromatography run at quite a low temperature. Condensatiion on the column of septum bleed and gas-line impurities for some time after injection can cause baseline liftoff and ghost peaks as the column is programmed up in temperature. The effect of these interferences is greatly reduced if the injector is switched to its split mode soon after injection.

Very large injections should be avoided with all types of columns and especially with capillary columns. The vaporized solvent produces a large volume of gas. Blowback and pressure-induced leaks just after injection are avoided by keeping sample sizes small. One- or two-microliter injections are both practical and reasonable. Capillary columns also have a limited capacity to handle large components and can be easily overloaded. However, it is usually better to dilute the sample rather than try to inject very small quantities. This is especially true if one wishes to take full advantage of the solvent effect. When a sample mixture contains components of widely differing concentrations, it is often necessary to run the sample a number of times at different dilutions if a complete analysis is required.

The Solvent Effect.   The large volume of sample and solvent vapor arising from an injection will sweep over a long length of cooled column before all the sample is condensed. This is equivalent to making a very long, slow injection. Much of the capillary performance will be lost, because all the sample will not start the chromatographic process from the same point.

Cooling the column below the boiling point of the solvent produces an interesting effect. The sample and the solvent will condense as a plug that completely fills the column. Movement down the column is very slow, normal injector pressures being insufficient to push the solvent plug very far. As the solvent evaporates from the front end of the plug, the less volatile sample components are concentrated in the remaining solvent (Fig. 103). Eventually all the solvent evaporates, leaving a narrow condensed ring of sample on the column. This solvent effect is analogous to zone refining.

Choosing the exact conditions in terms of injection amount, injector pressure, and column temperature will depend on the system and the solvent used; some experimentation is therefore necessary. Figures 104 and 105 illustrate the effect using isopropanol as a solvent. Isopropanol boils at about 80°C. An injection at 60°C produced excellent chromatography, whereas injections at 100°C gave poor results except for the relatively high boiling point components. These, of course, would condense very quickly in a small area at the head of the column even though it was relatively hot. Conditions for the two runs were similar. The

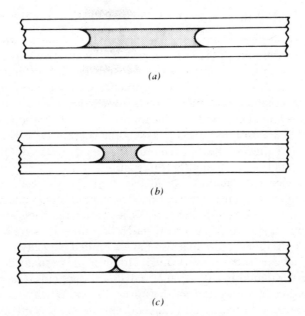

Figure 103.   The solvent effect in capillary column gas chromatography. (a) Long plug of sample in solvent. (b) Solvent evaporates, concentrating the sample. (c) A narrow starting band is formed when all the solvent evaporates.

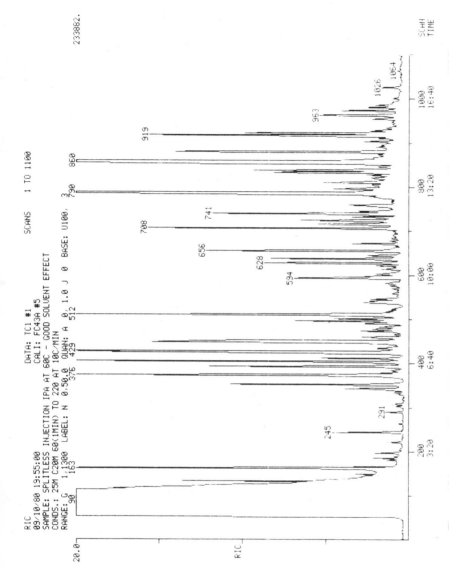

Figure 104. A capillary GC/MS sample in isopropanol injected with the column at 60°C (20°C below the solvent boiling point) shows a good solvent effect with excellent chromatography. Sharp peaks appear on the solvent tail even before scan 200. (By kind permission of Finnigan-MAT Ltd.)

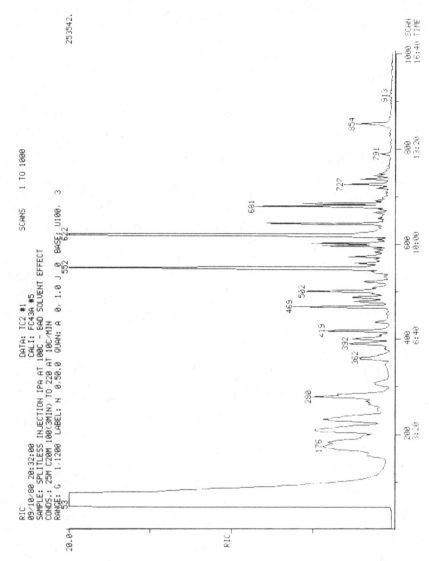

Figure 105. The same sample and conditions as in Figure 104, but with initial column temperature 100°. The solvent effect is bad, and the early peaks are unresolved. (By kind permission of Finnigan-MAT Ltd.)

better data were obtained with a 60°C initial temperature held for 1 min, while
the other data were obtained using a 100°C initial temperature held for 3 min.
The program rate of 10°C/min and all other parameters, including sample
amount and column, were the same. Comparison of the spectra in Figure 106
indicates that the peak at scan number 789 from the good data (upper spectrum)
is the same compound as in scan number 552 from the other run (lower spectrum). This shows that a lot of early information is lost with the hotter injection.

Although not entirely obvious from the RIC traces reproduced here, the 60°C
initial temperature produced slightly sharper chromatography peaks, giving a
slightly improved sensitivity (about 21% increase for the spectra shown). Hence
it is worthwhile experimenting to get a good solvent effect. Operating with an
initial temperature of about 20°C below the solvent boiling point is a good starting point. If 20°C is too cool to be practical, a higher boiling point solvent could
be used, but this is unsuitable if the sample contains very volatile components.
Injecting a greater amount of a lower boiling point solvent can sometimes help in
this situation. This is done by either diluting the sample and drawing up a larger
amount or by taking up some pure solvent into the syringe with the sample. Care
is needed so that the excess solvent does not damage the column.

The solvent plug should last for only about a minute for a good starting band
to be formed. To prevent respreading of the narrow starting band as well as to
prevent ghost peaks, the split valves should be opened at or just before the solvent plug finally evaporates. For example, opening the septum sweep and body
purge lines of the Grob injector illustrated in Figure 64 about 30 s after injection
with a solvent plug lasting up to a minute should prevent further solvent condensation and spreading of the sample. Much depends on the dimensions of the
column and the nature of the injector as well as on the solvent used. The operator should experiment to find the optimum conditions.

One criticism of this technique is that the column oven temperature program
must start from a low temperature. Although analysis time can be reduced by
multiramp programs, the constant cycling of oven temperature can be a cause of
leaks. Ferrules and fittings loose their seals with the repeated rapid expansions
and contractions. If the sample components of interest elute at higher temperatures, this problem can be overcome by using a higher boiling point solvent.
Figure 107 illustrates the excellent chromatography at very low levels and short
analysis time that can be achieved with this technique. Samples prepared from
horse urine extracts were made using dodecane as a solvent. Dodecane boils at
214°C. The samples were injected at 190°C. The temperature was held for 1 min
and then programmed up to 230°C at 20°C/min and then up to 265°C at 2°C/
min. The main component of interest (Fig. 108) eluted after 5.6 min compared
with 13.5 min when hexane was used as a solvent in the splitless mode. A retention time of 4.2 min was achieved using hexane in the split mode, but the sample
spectrum was too weak to show the low-intensity molecular ion at 422 u. Examination of the reconstructed gas chromatogram and the 422-u mass chromatogram shows that the compound of interest is only a small peak close to a major

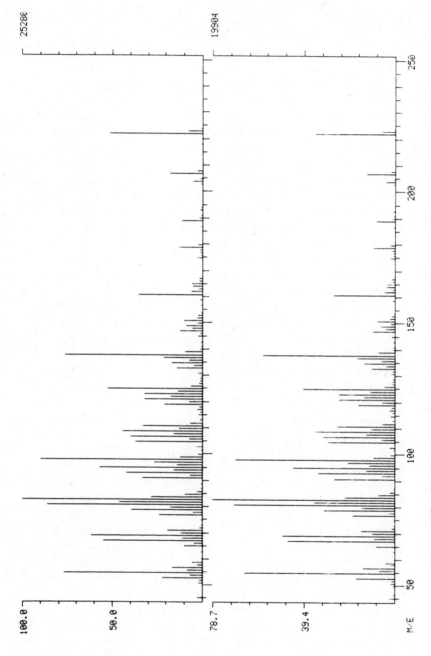

Figure 106. Equivalent spectra obtained from the RIC traces of Figures 104 and 105. The upper spectrum is scan 789 from the good solvent effect data. The lower spectrum is scan 552 from the bad solvent effect data and is about 21% less intense. (By kind permission of Finnigan-MAT Ltd.)

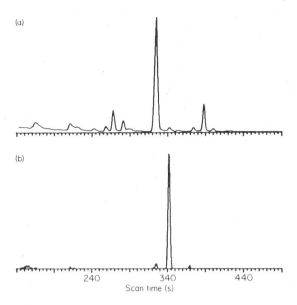

Figure 107. (*a*) Reconstructed gas chromatogram (mass range 100 to 450 u) of horse urine extract. Dodecane was used as a solvent, and splitless injections were made with an initial capillary column temperature of 190°C. (*b*) Mass chromatogram 242 u to detect estrane-3,17 α-diol bis-TMS. (From E. Houghton and P. Teale, "Capillary column gas chromatographic mass spectrometric analysis of anabolic steroid residues using splitless injections made at elevated temperatures," *Biomedical Mass spectrometry*, vol. 8, no. 8, 1981. © Heyden & Son Ltd., 1981. Reproduced by permission of John Wiley & Sons, Ltd.)

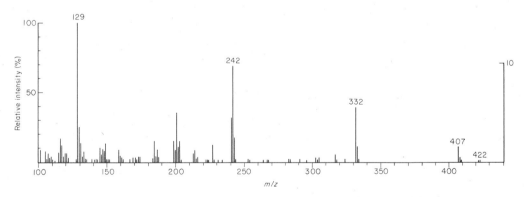

Figure 108. Mass spectrum of estrane-3,17α-diol bis-TMS. (From E. Houghton and P. Teale, "Capillary column gas chromatography mass spectrometric analysis of anabolic steriod residues using splitless injections made at elevated temperatures," *Biomedical Mass Spectrometry*, vol. 8, no. 8, 1981. © Heyden & Sons, Ltd., 1981. Reproduced by permission of John Wiley & Sons, Ltd.)

component and would be swamped without the high resolution available with a capillary column method.

On-Column Injection.   There are a number of disadvantages of the splitless injection methods using an injector. The cold sample and solvent are introduced into a hot injector body where evaporation takes place. To minimize contact of sample vapor with the hot injector body and to establish a satisfactory solvent plug, relatively slow injections are sometimes made. Evaporation is then predominantly from a hot needle surface, which may well introduce even more serious absorption or sample degradation. Using a heavy (higher boiling point) solvent and delaying the opening of the split valve may help to reduce molecular weight discrimination. The trade-off is that the very volatile components show poor chromatography, evaporating before the solvent in a nonuniform manner, and ghost peaks are more likely. On-column injection overcomes some of these problems.

In this method of injection the injector assembly is little more than a valve through which the sample syringe can pass (Figure 65). The syringe is guided through the valve into the column. The injector body is unheated and, in fact, may have additional cooling to prevent premature sample evaporation. Once in the column the syringe can be operated, leaving a cool solvent plug at the head of the column. The syringe is then withdrawn and the solvent effect allowed to act. There is no heating of the sample or intermediate vapor phase. A relatively volatile sample can be used, giving good chromatography even for components with low boiling points. Compared with splitless injections, chromatographic resolution just after the solvent tends to be better with on-column injection, but the technique does require additional care and attention if it is to be successful. The column ends must be flared so that the syringe needle enters easily. Injection of the sample onto the column must produce a single uniform plug of solvent. Care must be taken not to shoot the sample down the column or to smear it along the tube back into the injector if the full solvent effect is to be achieved.

Even if an on-column injector is not available, it is possible to obtain a similar result. A standard injector is used and operated cold. The column is disconnected from the injector and the sample in solvent is introduced into the end of the column with a syringe or fine pipet. The column is then reconnected to the injector and the conventional temperature sequence started. Good results can be obtained this way for very labile compounds. Obviously, the effort is not worthwhile except in extreme cases, but it is an option open to all capillary system users.

INJECTION TECHNIQUES

There is no single way to handle a syringe correctly, but there are many incorrect ways. Many good techniques are spoiled by carelessness or a simple oversight. Where much of the GC/MS work is qualitative, and quantitative work uses internal standards, many operators feel that a precise injection technique is of little importance. This is sad, because good syringe technique is fundamental to obtaining good and reliable results. All too often, poor chromatography is blamed on a bad injection and discounted. A result may be of poor quality but nevertheless valid. If the operator has confidence in his or her injection technique, then poor results are indicative of system problems. These can be identified and dealt with at an early stage. A couple of bad injections on a capillary system can cause a day of instrument time to be lost.

Even when the injection technique is adequate, lack of precision in the amount injected can lead to problems. On a manual system it is desirable that the major gas chromatography peak reaches almost full-scale deflection on the chart paper if the best dynamic range is to be achieved. For completely unknown samples, this is always a matter of guesswork, but on repeat samples 80 to 90% deflection for the major component should be routine. Although data systems allow one to manipulate displayed data, the argument is still a valid one. There are often instances where it is desirable to inject a certain minimum amount to detect small peaks yet care may be needed not to saturate another component or its spectrum. With some samples it is not always possible to achieve this in one injection, but for others it is a matter of precise technique.

On manual systems it is much easier to anticipate the correct starting point for a UV recorder scan if the peak deflection is known. On all systems great confidence can be placed on small peaks if they can be reliably repeated. Haphazard injections lead to traces of variable amplitude, making these things difficult. In many types of analysis it is necessary to run similar or even the same sample many times—for example, standards, control samples, and calibrations. A dependable technique should give equal TIC responses. Variations in the system response can be interpreted as changes in sensitivity. Many operators have found that this is a valuable way of identifying sensitivity problems.

For instance, instrument performance is a function of source cleanliness. Sensitivity slowly falls as the source becomes contaminated. Monitoring the response with routine samples enables one to anticipate necessary source-cleaning operations and in some circumstances can lead to considerable savings. For example, in a clinical trial it may not be possible to shut down the system for cleaning in the middle of a series of samples. Monitoring response factors in absolute terms by carefully injecting the same amounts of a standard each time allows the experienced operator to anticipate source lifetime. In this way a batch of samples that required many hours of work in preparation will not be wasted.

Syringe Handling

What are the important factors with regard to syringe handling? Probably the most important is a consistent method. In principle, no matter how "bad" the technique, consistency in method should lead to consistent results. Obviously, the better the technique, the more likely a consistent result will follow.

Try the following experiment, assuming that our intention is to inject 1 or 2 μL of sample. Take your favorite syringe (people's feelings and prejudices about their equipment are important, as they affect the way they use them) and draw up a microliter of solvent, say hexane. Empty the syringe onto a tissue; wipe the needle clean, taking care not to draw solvent out of it by capillary action. Pull the plunger up and observe the syringe barrel. Some syringes will still contain up to 0.5 μL of solvent, others will appear to have none, and on some types it is not possible to tell. The point is that there is a needle volume that may or may not be included in the calibration of the syringe volume as marked on the barrel.

The needle volume, of course, is boiled out of the syringe in a hot injector, but some solvent (and sample) may remain. For instance, try drawing up solvent to the l-μL mark on the syringe, injecting into a hot injector, holding it in the injector for the minimum time, and then withdrawing it. Pull the plunger up and see if any solvent remains. Often 0.1 or 0.2 μL of solvent is left. The solvent effect, operating in the syringe needle, may act to concentrate the less volatile samples into this small plug of solvent with only the pure solvent being boiled off. After a few injections the syringe may contain a relatively concentrated sample!

Consider the injection technique outlined below and the associated comments. This is not intended to serve as the last word in syringe handling but more as a basis for discussion. If it causes the reader to cast a critical eye over his or her own technique it will have fulfilled its purpose.

1. For a 1- or 2-μL injection, use a 10-μL syringe. The plunger is withdrawn only a small amount and is unlikely to be damaged. The operator does not have to have long fingers for a satisfactory technique.

2. Be sure the syringe is clean. Very thorough rinsing by pumping clean solvent through the syringe is recommended before and after use. Care should be taken not to touch the plunger shaft.

3. Have the sample, a beaker of clean solvent, and some tissues ready near but not on the instrument. Spilling solvent over the instrument can dissolve plastic switches, knobs, and fascias and could even start a fire.

4. Pump solvent through the syringe by drawing up 4 or 5 μL and jetting it onto a tissue. Do this a few times. Then pump the syringe with the needle held in the solvent. Draw up a few microliters and then empty the syringe in a smooth action onto a tissue. Wipe the needle clean, taking care not to draw out any solvent. This wets the inside of the syringe and fills the needle volume with pure solvent.

5. Transfer the syringe to the sample bottle without delay. Draw up the sample to the required mark. Always view the level from the same angle and use the same point on the graduation line. (The lines have a finite thickness.)

6. Remove the syringe from the sample, and wipe the outside of the needle, taking care not to draw out any sample. Pull the plunger up a further

1 μL to draw the sample into the syringe body. Most of the sample will now be held inside the syringe barrel and is unlikely to be lost due to spillage or evaporation. Observe the syringe barrel. A very small air bubble is often visible between the pure solvent and the sample, allowing the true amount of sample to be assessed. There will be an additional amount corresponding to the wet inside needle surface.

7. Inject the needle through the septum with a smooth movement. Hold a finger above the plunger to prevent it blowing out. For most methods a firm and quick injection is necessary. Leave the needle in the injector for only a few seconds. The greatest danger of sample loss occurs on injection. The hot injector may cause evaporation (hence pulling off the sample into the syringe barrel) before insertion is complete. It is important to get the needle into the column, not down the side.

   Observe the sample in the barrel and the air bubble described above. Blowback due to a leaking plunger should be noted and the syringe replaced. Do not pull the syringe out suddenly immediately after injection, or the samply may blow out through the needle hole in the septum. A 3- or 4-s delay is appropriate. Some capillary techniques require a slow injection so that the solvent is carried onto the column evenly and not blown down a long length. Modify the technique as necessary. A faster technique may be necessary with particularly labile compounds to prevent degradation on the hot needle surface. Blowout losses are the lesser of the two evils in these cases.

8. Withdraw the syringe and pump clean as described in step 4. This procedure should be carried out quickly and smoothly. Operation of diverter valves, starting temperature programmers, and data system acquisition or chart recorders should be incorporated into the procedure so that it is automatic, unhurried, and precise. Guard against distractions at this critical moment. Do not forget to close the diverter and/or switch on the filament as necessary. Some operators find that cleaning the syringe by flushing it out a fixed number of times is a good way to time these procedures and ensures a smooth rhythm to the operation.

Needle Guides. There are arguments for and against needle guides. On the one hand, they ensure that the needle enters the column correctly. The risk of accidental needle bending is reduced and guides are therefore advantageous with some types of syringe. On the other hand, the needle always punctures the septum at the same spot, and septum life may be reduced as a consequence. Needle guides can also transfer more heat to the syringe barrel and so care should be taken not to press the syringe right down onto the guide.

Cleaning Syringes. When changing from one sample to another, it is important to clean the syringe thoroughly by rinsing it with pure solvent. From time to time it is desirable to remove all traces of organic material by rinsing through with a cold chromic acid solution. Most syringe manufacturers will supply instructions for cleaning their products.

Occasionally a small piece of septum will block the fine needle bore. On some designs the needle can be cleaned by disconnecting it, filling it with solvent, and then blowing out the blockage by reconnecting the needle to the body and plunger. Another way is to wet with solvent the area where the plunger enters the body. Completely withdrawing the plunger creates a partial vacuum in the syringe and the solvent will rush in. The plunger can then be reinserted and the blockage pushed out. Excessive pressure may damage the syringe, and it should be checked for sample blowback after clearing blockages in this way.

CHAPTER EIGHT

Mass Spectrometer
Operation

The mass spectrometer in most GC/MS systems may be operated in a number of different modes depending on the required type of analysis. Within any particular mode, the operator usually has many degrees of freedom in terms of instrument parameter selection. That is to say, the operator usually has the choice of various compromises necessary for a successful analysis. That compromise is necessary is beyond question; the secret is to obtain the best overall result.

As before, we must limit our discussion to factors relating to the choice of methods and to the operator decisions made once a method has been chosen. Some comments about the operating characteristics of different types of mass spectrometer are relevant, but these may not always be general. Certain features may be particular to only one model of instrument and not applicable to others of a similar type. The reason for this is that some of the decisions that must be made have already been determined by the manufacturer of the system and the operator has a reduced range of choices. For instance, the GC/MS system may be an EI-only instrument and so different ionization modes cannot be selected. However, for the purposes of our discussion we will assume that the widest range of choices is available, in which case ionization mode is probably the first major mass spectometer choice to be made when starting an analysis.

IONIZATION MODES

The selection of source mode will depend on the type of sample being analyzed. In general, unknown samples are probably best run under EI conditions. This will yield the maximum amount of structural information and provide spectra for library searching if the facility is available. Even without computerized library search routines, there are a number of peak index publications that can help with classification. Softer ionization techniques such as chemical ionization provide more particular information about a component and are ideally suited to the detection of known compounds.

In the study of unknowns it is usual to combine the structural information obtained in EI with the molecular weight and functional group information that can be obtained with CI. Some mass spectrometers can be switched from one mode to another during a run, using alternate CI/EI (ACE). If not, the sample must be run twice or with a high level of helium carrier in the source to give mixed EI and CI due to a charge-exchange spectrum as well as the normal CI spectrum.

Electron Impact Ionization

Although EI spectra provide a lot of information about the sample components, there are some drawbacks. Fragmentation of some compounds is so great that only weak molecular ions may be seen and in some cases they may be missing altogether. Often the major ions are well down the mass range and subject to significant interference from the background spectrum. Background-subtraction and data-enchancement techniques are an obvious advantage in these cases.

On manual systems a UV recording of the background spectrum taken just after component elution does not always provide the best background reference. It would appear that eluting samples often carry phase and impurities out of the column with them and, of course, any chromatography peak trailing will lead to some sample ions being present in the background UV trace. It would be better to take a trace just before sample elution, but this implies that the sample retention time is accurately known.

If the sample has been run previously, either on the GC/MS system or on a stand-alone gas chromatograph, this information, or at least an approximation, can be deduced from the trace obtained. This is one good reason for running all samples on a stand-alone gas chromatograph first. Even so, these traces do not always translate to the GC/MS or, for a number of reasons, they may be unavailable. In these cases it is best to take a few recordings of the background during the run at "flat" times. If the column oven is programmed up in temperature, more traces should be recorded at frequent intervals to allow for any change in bleed levels. In this way a realistic estimate of the true background ions and their levels can be made. One should always be prepared for sample ions at the same masses as the background ions. Background subtraction in these cases reduces the intensity but does not completely eliminate the ion in question.

Another problem associated with EI spectra is the high level of noise (sometimes called "grass," due to its appearance in the spectrum) caused by the excited and ionized carrier gas. To overcome this problem it is quite common to run the ion source with an electron energy voltage below the ionization potential of the carrier gas. The ionization potential of helium, the usual carrier gas in EI

GC/MS work, is 24.6 V, and so operation at an electron energy voltage of about 20 V is common. Noise from the carrier gas if then greatly reduced. Total ionization of the sample is also reduced, leading to lower sensitivity, but the detection limits may be improved.

This apparent paradox is easily explained. Fragmentation of the sample is reduced by the lower energies so that although there is less total ionization current there is a larger proportion of it in the higher mass ions. The gas chromatography peak is observed in the presence of less noise and may be more easily seen. Its spectra will contain stronger high-mass ions away from the majority of background signal and they will be subject to less grass. Signal-to-noise ratios and therefore detection limits can consequently be improved for some compounds.

Operation at low electron energies leads to spectral cracking patterns that vary greatly, especially if other factors such as source temperature are not carefully controlled. It is for this reason that the majority of EI work has been done at 70 eV. Spectral cracking patterns at this relatively high energy do not vary so much and are more dependable when used for comparison purposes, the reference spectra often coming from another instrument or even a different make of instrument. This does not mean that EI spectra for library searching have to be run at 70 eV. In fact, some instruments will tune for maximum sensitivity at 40 to 50 eV, but cracking pattern ratios may not change until the electron energy drops much closer to the ionization potential of the carrier gas. The electron energy voltage setting should be investigated during source tuning along with the other source variables to achieve the best overall response.

Chemical Ionization

Probably the second most common ionization mode is chemical ionization. (CI). In general CI leads to simpler spectra, with a few ions at the higher end of the mass range. In many cases these ions are directly related to the sample's molecular ion and are known as quasi- or pseudo-molecular ions. The degree of cracking and adduct ions depend on both the sample and the reagent gas used.

Certain classes of compounds yield strong negative ion spectra when analyzed with a CI method (Fig. 109). On some instruments the operator must decide beforehand whether to run in the positive-ion or negative-ion mode. On others the instrument can be quickly switched over or even operated in a dual mode, collecting both positive-ion and negative-ion signals. Selection of ion polarity and the choice of the best reagent gas implies that there is some knowledge of the sample chemistry. For instance, it is generally true that compounds that give strong signals with electron capture detectors will also give strong negative-ion spectra.

Some reagent gases are selective in the way in which they attack the sample compound and can yield structural information, providing that the sample is a compatible compound. Compounds of different chemical classes may not even be ionized. Hence, CI can provide a means of specific detection for compound class and distinguish between structural or isomeric variations, but probably its

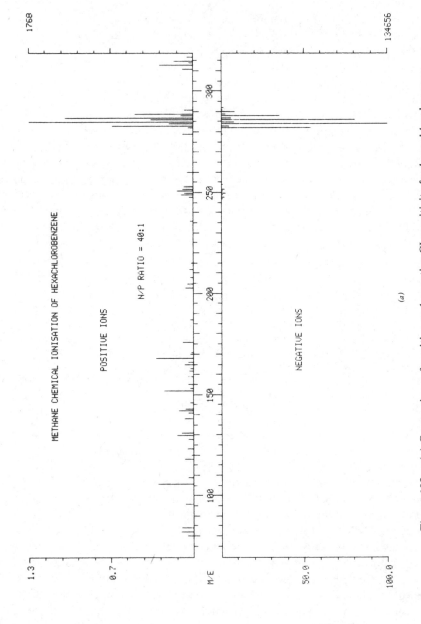

Figure 109. (*a*) Comparison of positive and negative CI sensitivity for hexachlorobenzene. The negative-ion to positive-ion ratio is about 40 to 1.

190

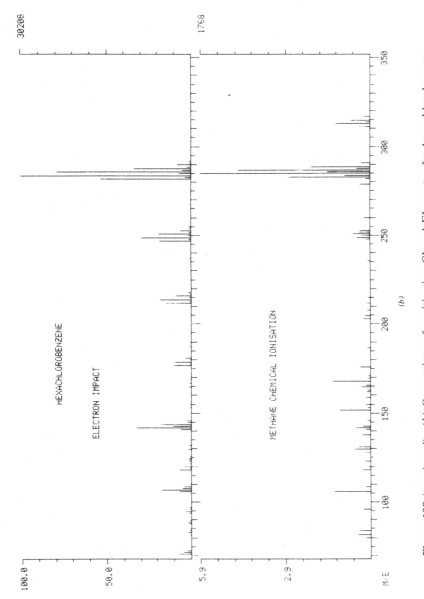

Figure 109 (*continued*)   (*b*) Comparison of positive-ion CI and EI spectra for hexachlorobenzene. The EI spectra is about 15 times more sensitive than the CI. Hence negative-ion CI (methane) gives the best overall sensitivity. (By kind permission of Finnigan-MAT Ltd.)

greatest use is in providing strong ions at the molecular or quasi-molecular weight (from which the molecular weight can then be determined). In this application details of the sample may be vague, and so a general-purpose CI reagent gas must be chosen.

Methane is probably the most popular CI reagent gas, but ammonia is also quite often used. Some companies in the past seemed to favor isobutane as a routine CI reagent gas in their instruments. The unkind implied that this was because their systems were unable to pump the lighter reagent gases efficiently at normal CI pressures. A more likely reason is that isobutane has a greater proton affinity than methane and will give less fragmentation. In the early days of CI it was considered desirable to produce sample spectra with as few ions as possible. This is a valid argument even today, and the use of isobutane as a reagent gas in some applications is practical as well as traditional. However, there is a danger that there will be no ionization at all, let alone fragmentation, and so methane with a lower proton affinity is more reliable as a general-purpose CI reagent gas for use with unknown samples.

Although the remarks leveled at some manufacturers with regard to isobutane were unfair, the gas does pose some problems for all GC/MS users. It is a relatively heavy gas and gives a reagent gas spectrum that extends well up the mass range. This background interference can mask important sample ions. Furthermore, isobutane is not kind to hot filaments. Source and ion gauge filaments tend to have a relatively short life when used with it. Of course, this is a general problem with all CI work, and the effects of different gases varies from instrument to instrument. Ammonia has a relatively high proton affinity and gives little fragmentation when used as a reagent gas. It gives a much cleaner reagent gas spectrum and has as a consequence become popular in recent years.

These general-purpose CI gases yield a large number of slow or thermal electrons as a result of their ionization as well as the protonating reagent ions. Compounds with a tendency to electron capture will pick these up, giving negative-ion spectra. Methane, in particular, is used as a general-purpose negative CI reagent gas. However, its action does depend on the sample (Fig. 110). For general-purpose negative CI work with a wider range of compounds, nitrous oxide mixed with methane has been used successfully. The reagent plasma contains a high density of OH- ions, which generate negative-ion spectra by hydride extraction from the sample molecule. Nitrous oxide is unfortunately a vigorous oxidizing agent, and filament life is shortened. Another problem is inadequate mixing of the methane with the nitrous oxide. This can be overcome by using methane as the gas chromatography carrier gas and adding the nitrous oxide as the CI reagent gas through the normal CI gas inlets.

CI Reagent Gas Routing.   Another important aspect of CI that must be considered is the routing of the reagent gas. Ideally the reagent gas should be used as the carrier gas so that all the sample is carried into the ion source. In packed column work with methane as carrier/reagent gas this is practical, but for capillary work and most other reagent gases it is not. In these cases helium must be

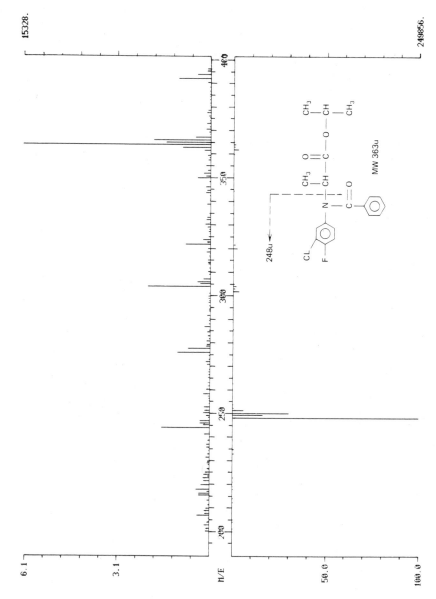

Figure 110. Methane CI of the herbicide flamprop—isopropyl. The positive ion spctrum yields $M+1$ and $M+29$ ions at 364 and 392 u as a result of addition reactions. The negative-ion spectrum shows a very strong ion at 248 u due to electron capture and dissociation at the nitrogen atom.

193

used, with the reagent gas added into the ion source. The high pressure of helium in the source tends to distort the CI spectra, causing additional fragmentation by charge exchange. Gas flows from capillary columns are not too great, but packed column work poses a problem. Direct connection to the ion source gives maximum sample utilization but a high level of "helium CI." The use of separators overcomes this problem but their efficiency is impaired by the abnormal outlet (source) pressure compared to EI operation for which they were probably designed. As a result, sample loss may be high.

Once again it is up to the operator to make a decision as to the best method for the particular system. Experimentation with the different options is necessary. In all cases care must be taken to ensure that the reagent gas does not blow the sample out of the ion source, or to one side of it, preventing adequate ionization. Reagent gas is usually added as makeup from the same side of the ion source as the gas chromatograph, but one should not automatically assume this gives the best results. It is worth experimenting if other reagent gas inlets are available.

TUNNING AND CALIBRATION

Tuning the Mass Spectrometer

Before any data are acquired it is necessary that the mass spectrometer be tuned satisfactorily. Source pressure is a significant factor affecting tuning, and so it is desirable to tune the instrument under normal operating conditions. Introducing calibration or reference compounds for tuning upsets the normal source pressures and may lead to non-optimum tuning parameters. Furthermore, these compounds tend to contaminate the ion source relatively quickly and should be used for as short a time as possible. Under normal operating conditions there are nearly always some background or bled ions that can serve for basic tuning purposes. Certainly up to about 300 u ions should be visible in a sensitive system. If not, then the instrument is not operating at its best.

Above 300 u there are few background or bleed peaks, even on a very sensitive instrument. The introduction of small amounts of reference material is then necessary. In fact many operational protocols insist on it. Opening up source apertures, reducing column flow into the mass spectrometer, and restricting the amount of the reference compound help to reduce tuning variations due to pressure changes. In this way, reproducible tuning over a wide mass range is possible.

The mass spectrometer should be tuned for maximum sensitivity consistent with good mass spectral peak shapes high-to-low mass ratios, and required mass separation (resolution). Good (that is, symmetrical), peak shape is important if accurate mass assignments are to be achieved. Even for nominal mass work, mass measurement to 0.1 u is desirable. Peaks that are skewed or ragged in appearance make it difficult for the operator to determine the center of mass. Data system mass assignments are dependent on the quality of the signals provided, and one should guard against overconfidence in a computer's ability to sort out poor information. For those data systems that scan the mass spectrometer by jumping from mass to mass, there is always the possibility of reduced sensitivity because of a local minimum in the peak profile. Smooth, round-topped, if possible flat-topped, mass peaks minimize this problem.

Some of the tuning parameters affect mass peak width at baseline as well as sensitivity and shape. This appears to change instrument resolution. On quadrupole instruments, for instance, liftoff at the front and rear of mass peaks can be caused by the ion energy being too high. (Ion energy is called quad. offset or draw-out on some systems.) Better peak shape is often obtained by running at a lower setting of ion energy. If the peaks are then over-resolved, the resolution controls should be adjusted. The base width of the peak can then be restored, giving more signal and a symmetrical peak shape and resulting in better overall performance.

A number of tuning parameters are mass dependent; that is, their value is made to change as the mass spectrometer is scanned. Ion energy programming on quadrupole systems is a good example. The ion energy may be optimized at one mass but may not be suitably adjusted at higher or lower masses. Where a parameter has two controls, a fixed value and a mass-dependent (programmed) value, it is advisable to turn off the programming control until the other parameters are adjusted. Each time the mass range is altered, the parameter, ion energy in this example, should be reoptimized. The necessary adjustments are a good indicator of the amount of programming that will eventually be required. The program value can then be set up after the other parameters have been optimized. This technique minimizes confusions caused by unknown programming rates and reduces the likelihood of crossed controls.

This problem of *crossed controls*, one parameter set too far one way and compensated by another also set at a nonoptimum value, is quite common. Tuning or at least fine-tunning is usually achieved by small adjustments of the controls. To check for local optima, which may not give the best overall performance, each control should be adjusted boldly across a wider range. Sometimes setting one control away from an apparent optimum and then readjusting the others leads to a better result. This of course sounds like guesswork and to a certain extent it is. However, the operator should experiment with the instrument and get a "feel" for the effect of each control. With experience one can usually adjust the controls to a set of values and be confident of being close to optimum tuning conditions.

A detailed log of settings is invaluable, since tuning parameters under a wide range of source conditions can then be established. Abnormal settings indicate problems due to either contamination, poor connections, or electronic malfunction and should always be investigated. Bold changes of a parameter are also quite a good way to identify charging effects. If the spectrum response is sluggish or tends to be rapid and followed by a slow drift to a new value, then charging in the source or analyzer should be suspected. Similarly, if there is no response at all to a change, then the connections, or lack of them, to the particular lens plate must be considered.

Automatic tuning routines under computer control are becoming increasingly available. In the early days of computer-controlled tuning it was said by some that if the system were basically in tune then the computer would not mess it up too much! Computerized tuning routines have improved considerably, but their use should be considered carefully and the resultant tuning should always be checked. There are two basic reasons for this attitude. First, the routines always rely on the introduction of a reference gas, which is usually a fluorinated hydrocarbon. As discussed above, this changes the source pressure and contaminates the system. Second, while control of the mass spectrometer parameters by a computer is relatively straightforward, the necessary algorithm for deciding on the correct set of values is not. Operator intuition cannot be written into the tuning program. The computer program also must assume certain starting conditions or at least a limited range of conditions, whereas a skilled operator will quickly identify and correct unusual but not necessarily faulty conditions. Automatic routines are improving all the time and will eventually be entirely satisfactory. Until then the operator is cautioned to monitor results and to recognize that automatic tuning routines are a convenience offered at the price of reduced source lifetime.

Major instrument tuning should be done only after a significant change of instrument status, after source cleaning for instance. Minor adjustments should be made as necessary; certainly the system should be checked at the start of each day's work. Some parameters do not have to be adjusted every day but should be checked at reasonable intervals. Periodic checking of rf generator dip or matching on quadrupole systems and resolution settings on magnetic instruments are typical examples of this type of adjustment.

Tuning and calibration should not be confused. The purpose of tuning is to achieve the best mass spectrometer sensitivity across the mass range consistent with peak shapes that enable mass measurement and possibly quantitation to the required degree of precision. The choice of word *precision* is deliberate. For reliable work, mass setting and measurement must be repeatable on a scan-to-scan, run-to-run basis. Adjustment of parameters that also affect mass measurement should be kept to a minimum and the system recalibrated afterwards. Obviously, there are different levels of adjustment. For instance, it is reasonable to increase resolution to give slightly better mass separation so that there is no cross-talk between isotope peaks when dong quantitative work. The change from the settings for normal maximum detection, unity mass assignment may be small, and recalibration may not be necessary. It should be noted that changing resolution can also alter the mass peak positions, and larger changes or a requirement to operate with accurate mass assignments will necessitate recalibration of the system.

DFTPP Tuning. Many manufacturers now provide routines that will tune their instruments to match the fragmentation pattern of DFTPP (decafluorotriphenylphosphine) (Fig. 111); as described by Eichelberger, Harris, and Budde.* This tuning pattern has been adopted as a standard for instrument

*J. Eichelberger, L. Harris, and W. Budde, *Anal. Chem.*, **47** (7), 995 (1975).

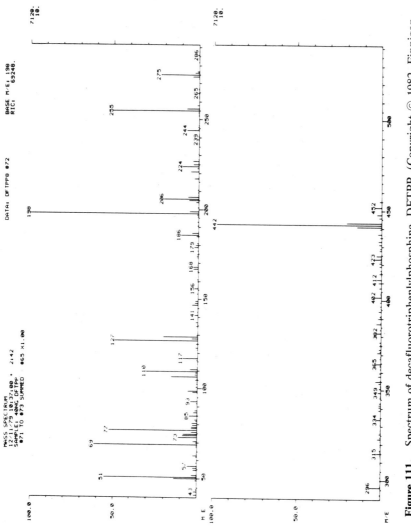

Figure 111. Spectrum of decafluorotriphenlylphosphine, DFTPP. (Copyright © 1982, Finnigan Corporation. All rights reserved)

performance in U.S. priority pollutants legislation. While this legislation is not effective outside the United States, the protocols described are being adopted elsewhere, and so DFTPP tuning is becoming more common. DFTPP is usually introduced via the gas chromatograph as a GC/MS sample and will therefore indicate overall system performance under normal operating conditions.

Unfortunately, tuning on an eluting peak is difficult at best and does not allow much time for adjustments. To overcome this, manufacturers give tuning ratios for calibration compounds such as FC43 that should give equivalent DFTPP tuning. Of course, the compound must be run as a sample to confirm this finally. There are two main approaches to achieving the required ratios. One method involves modifying the source tuning of the instrument so that the necessary ratios are obtained. This changes the characteristics of the mass spectrometer, and an appropriate calibration routine is usually provided. The operator should watch out for distorted or unstable spectra due to extremes of tuning when using this method.

An alternative approach involves tuning the instrument to achieve the best overall performance. An abundance normalization factor—that is, a conversion graph—is then calculated so that all acquired spectra can be converted to a DFTPP format. The operator then has the option of displaying data in a converted or unconverted mode. Calibration routines for DFTPP tuning normally give output displays (Fig. 112) that show the conversion graph. The operator should watch out for large or sudden changes in the conversion factor. These indicate poor basic calibration and tuning of the instrument and can lead to very distorted displayed spectra.

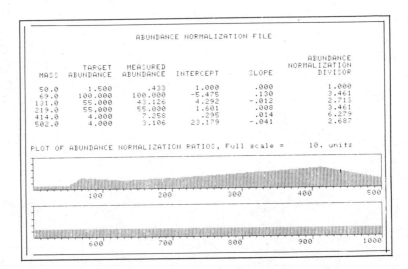

Figure 112. Abundance normalization display following tuning. Used to adjust displayed ion intensities to satisfy DFTPP tuning protocols. (Hewlett-Packard Corporation)

Calibrating the Mass Spectrometer

The assignment of mass values to acquired mass spectrometer data requires that the instrument be calibrated. The degree of accuracy required will determine both the frequency with which calibrations must be made and the way in which they are achieved.

Mass Markers. For real-time mass displays (spectrum monitors), mass markers are usually calibrated by adjusting the marker's position and its readout to agree with two reference peaks at either end of the instrument mass range. Intermediate positions are checked, and the two set points are adjusted to give an acceptable tolerance across the whole spectrum. On nonlinear sector instruments, more set points may be used so that a series of reference points are established. Nominal mass accuracy is easily obtained with these techniques, and accuracies of 0.1 u are often possible. Once set, the mass marker should be checked regularly against known background peaks and adjusted as necessary.

Monitor mass markers are usually a single bright up spot, blip, or cursor which the operator can move across the displayed spectrum. Often associated with this single mass marker is a continuous or multiple mass marker that is used to mark UV recordings of spectra.. This too should be regularly calibrated. The fine detail available from the continuous mass markers is often difficult to see when superimposed on the monitor display. To overcome this problem the following calibration technique has been found useful.

The continuous mass marker and the single mass marker can often be displayed together. Once the single mass marker has been calibrated there is no further need to display spectra. The calibration gas, and the ion source filament and multiplier can be switched off. The mass markers are now the only signals visible on the monitor, and the continuous marker can be adjusted to agree with the single mass marker. This technique greatly reduces both the time spent on calibration and the amount of reference compound used.

Data System Mass Calibration. The methods used fall into a number of categories. In all cases a calibration compound must be introduced into the mass spectrometer. In the first category the data system controls the mass spectrometer mass set input and searches for the mass peaks of the reference compound. Finnigan's 6000 series data system, for instance, jumps from mass area to mass area and then hunts to find the maximum signal intensity. This is logged in as the set point corresponding to the reference mass. The system must have information so that it knows where to look to find the reference peaks. A standard set of scan and window voltages are provided for the initial hunt. Once the system has been calibrated, the standard set of voltages can be upgraded to the last calibration set. Providing there has been no major change to the instrument or to the data system, subsequent calibrations are quickly obtained. This type of calibration is only suitable for nominal mass work; that is, an accuracy of about ±0.1 u can be obtained, which is sufficient for basic mass defect compensation.

Greater accuracy is obtained by locating mass peak centroids. This compensates for small variations in peak shape and the effects of system electronics and ion statistic noise. The mass spectrometer is scanned, not necessarily but usually, under computer control. Each peak is identified (defined as a peak) and assigned an intensity corresponding to its integrated area and a position relative to the start of the scan. The position parameter may be a time or scan output voltage. More than one scan of data can be acquired this way. Calibration consists of matching the recorded spectra to a calibration reference table. A number of different algorithms are available.

Different companies use different programming techniques, and any one manufacturer may offer a range of methods depending on the required accuracy. The principles, however, are essentially the same. The recorded spectrum is searched to find the most intense mass peaks. From the reference table and the scanning information their approximate position can be calculated. Once found, the base peaks can be used to give an initial calibration curve. This curve can be used to locate the smaller peaks in the reference table. Each time a peak is located it can be used to update the initial calibration curve.

Projecting up and down the mass range from each reference data point should locate the next reference mass peak. The system can use this technique to check for false assignments. The number of reference peaks used and the curve-fitting mathematics control the accuracy of the final calibration. Figure 113 shows typical outputs from a Finnigan-MAT Incos data system connected to a model 4000 quadrupole mass spectrometer and a model CH5 sector instrument. The computer outputs show not only the calibration fit error over the selected mass range but also other graphs relating to instrument performance.

The jagged appearance of these graphs is typical of data system calibration outputs. The graphs are normalized to a given display size and the axes scaled accordingly. The computer then draws straight lines between the points defined by the location of the calibration compound peaks. Accuracies of a few millimass units even on low-resolution data are routinely possible with techniques such as this. For more accurate measurements, high-resolution and/or peak-matching techniques must be used.

Peak Matching.   This type of mass assignment is achieved by comparing sample peaks with those of a reference compound. The comparison process realizes a ratio measurement that can be used to calculate the unknown mass. The reference compound must be present in the mass spectrometer at the same time as the sample. This may not be practical under GC/MS conditions. The reference compound is introduced into the source while running under high resolution conditions. High resolution is necessary to separate the sample mass peaks from the calibration gas ions.

An alternative approach for lower resolution work developed by AEI (now Kratos) uses a dual-beam technique. The system is basically two superimposed sector mass spectrometers sharing the same control voltages and fields. Both ion beams will be subject to the same variations. The sample is introduced into one side of the source, and a calibration compound is ionized in the other side, giving both signal and reference beams. Simultaneous acquisition of two beams must be made so that the mass positions can be related. This is a practical arrangement for GC/MS work.

Switched positive-ion, negative-ion operating modes for quadrupole systems using a reference compound with a predominantly negative-ion spectrum and the sample in the positive-ion mode have been used to achieve similar results. In commercial switched systems, positive- and negative-ion scans are acquired alternately. Scan-to-scan drift is usually small, and mass accuracies of a few parts per million have been obtained. Unfortunately, the sample and reference spectra are not completely independent. The calibration gas will give some peaks in the positive-ion spectra, and the sample may give negative ions also. The arrangement does not have the benefit of high resolution to separate these masses if they happen to interfere.

Signal Levels. Calibration, like tuning, should be done under as near to normal operating conditions as possible. This can lead to problems of detection of ions over a wide mass range. The ratios of the high-mass ions to the base peaks are often very small. One is often faced with the problems of insufficient signal or saturated base peaks. One remedy is to limit the calibration to the mass range of interest. In general terms this policy applies to tuning as well.

There is little to be gained in getting everything adjusted exactly up to 500 or 600 u if all your work is to be done on compounds of low molecular weight. The tuned and calibrated mass range should normally be set for the range of interest with a reasonable margin to allow for the unexpected. This does not mean that one can afford to adopt a casual attitude toward tuning. For instance, generator adjustment on quadrupole instruments becomes more critical at higher mass where the power levels are greater. Just because the instrument is to be operated at fairly low mass does not mean that rf tuning should be ignored. Poorly tuned rf generators dissipate more power at any given mass and will tend to overheat. This can lead to significant drift even at low mass.

Another way to deal with signal dynamic range on systems that jump and search for calibration peaks progressively across the mass range is to start with the multiplier voltage set low so that the base peaks are not saturated. As the instrument jumps to higher masses, the multiplier voltage can be wound up, giving a stronger signal. Obviously, this technique will not work so well for calibration compounds with base peaks in the middle or at the high end of the mass range. It also presupposes that the operator can tell the approximate mass setting of the mass spectrometer at any given time during the calibration procedure. This is not always possible.

For systems that acquire data and then compare them with a reference table, the problem of wide signal intensities can be minimized by acquiring several scans of data. During the acquisition the signal intensity is slowly reduced either by turning down the multiplier voltage or better still by reducing the amount of reference compound. This can often be done simply by switching off the refer-

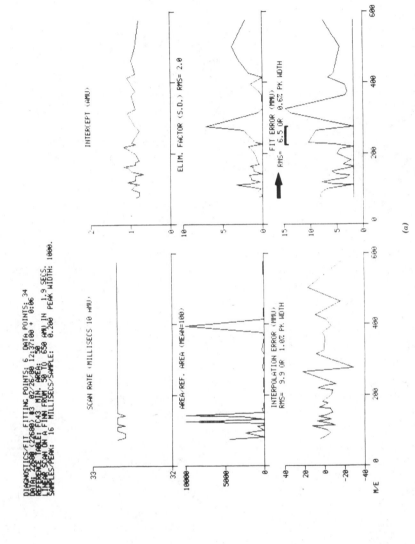

DIAGNOSTICS/FIT FITTING POINTS: 6 DATA POINTS: 34
DATA: 22680 (22680) 83 02/26/80 12:37:00 + 0:06
REFERENCE TABLE: FC43 MIN. AREA: 50
LINEAR SCAN ON A FINN FROM 50 TO 650 AMU IN 1.9 SECS.
SAMPLES/PEAK: 16 MILLISECS/SAMPLE: 8.200 PEAK WIDTH: 1000.

INTERCEPT (AMU)

ELIM. FACTOR (S.D.) RMS= 2.0

FIT ERROR (MMU)
RMS= 6.5 OR 0.62% PK WIDTH

SCAN RATE (MILLISECS 10 AMU)

AREA-REF. AREA (MEAN=100)

INTERPOLATION ERROR (MMU)
RMS= 9.9 OR 1.02% PK WIDTH

M/E

(a)

Figure 113. Calibration diagnostic Printouts. (*a*) Typical quadrupole instrument. (Copyright © 1982 Finnigan Corporation. All rights reserved.)

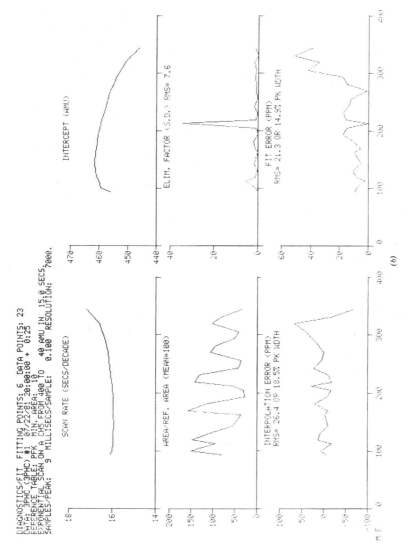

Figure 113 (*continued*) (*b*) Typical magnetic sector instrument. (Department of Chemistry, University College, Cardiff)

Both sets of data show computer-generated calibrations based on standard calibration tables stored in memory and on the acquired data using the standard compound. This was FC43 for the quadrupole and PFK for the sector instrument. Quadrupole parameters are scaled in milli-mass units for a nominal peak width of 1000 mmu. The sector instrument parameters are scaled in parts per million with a resolution figure of 7000.

ence inlet and allowing the compound to pump away. The signal decay should be slow compared with the scan speed so that each spectrum is taken at a reasonably steady level. Calibration can then be attempted on a number of scans. Ideally, one should use a scan without saturated peaks but with sufficiently strong minor ions across the mass range of interest.

Calibration Compounds

Reference compounds for mass calibration should give spectra with ions spaced evenly across the mass range of interest. The range of ion intensities should be small, certainly small enough to be used satisfactorily on a single setting of the mass spectrometer signal gain. The compound should be easy to introduce into the ion source and cause no major contamination problems. Although a number of compounds are used routinely, none fully satisfies the criteria above, but even so with careful use good calibrations are possible.

FC43 (Perfluorotributylamine).   This is also known as PFTBA and is used extensively by manufacturers of quadrupole instruments as a calibration compound. It is a liquid at room temperatures and is sufficiently volatile for direct introduction into the mass spectrometer. In fact, it is normal to introduce the vapor through a restrictor to limit the flow. Figure 114 gives a listing of the major ion fragments and expected intensities on a quadrupole instrument operated in the EI mode, and Figure 115 presents a typical EI spectrum. FC43 gives significant peaks in both positive-ion and negative-ion CI (methane), as illustrated in Figure 116.

   The upper limit of 614 u in the EI mode is sometimes a drawback, as is the relatively low intensity of ions in the 300 to 400 u mass range. It is an oily substance and contaminates ion sources relatively quickly. As with most calibration compounds it should be used sparingly.

PFK (Perfluoronated Kerosene).   This has a larger mass range than FC43 and regularly spaced peaks. It has been used extensively by manufacturers of magnetic instruments, especially for peak-matching work. It is less volatile than FC43 and must be introduced through a heated inlet system. PFK also gives regularly spaced peaks in the 200 to 600 u range when ionized in the negative-ion CI mode. Figures 117 and 118 give listings with typical quadrupole relative intensities for both the EI positive-ion and CI negative-ion modes.

Triazines.   Tris(perfluoroheptyl)-s-triazine has ions to 1185 u and is therefore extremely useful for calibrating instruments to a relatively high mass. It is not a very volatile compound and must be heated to be introduced into the ion source. A reasonable way to do this is to put a small amount in a glass bulb, which is then connected to the GC/MS interface inside the GC oven in place of a column. The GC oven can be carefully heated until a sufficient level of the compound is present in the source (Fig. 119). As with FC43 and PFK, the rate of source con-

```
INPUT FILE: FC43
OUTPUT FILE: FC43
   MASS     INTENSITY
  18.0106        1.
  19.9812        1.
  28.0061        1.
  30.9984        1.
  39.9624        1.
  43.9898        1.
  49.9968      312.
  68.9952   209920.
  75.9999        6.
  80.9952      562.
  92.9952       37.
  95.9983      375.
  99.9936     7760.
 106.9983        1.
 111.9936       52.
 113.9967     3732.
 118.9920    10192.
 123.9936        1.
 125.9967        1.
 130.9920    63168.
 137.9967        6.
 142.9920        1.
 149.9904     1566.
 154.9920        1.
 161.9904       52.
 163.9935     1144.
 168.9888     4440.
 175.9935      715.
 180.9888     2368.
 187.9935       71.
 199.9872      195.
 206.9919        1.
 213.9903      936.
 218.9856    86016.
 225.9903      558.
 230.9856      570.
 237.9903        7.
 242.9856        5.
 256.9887        1.
 263.9871    16768.
 275.9871       63.
 287.9871        1.
 294.9855       20.
 299.9871        1.
 313.9839      497.
 325.9839      118.
 337.9839       18.
 344.9823        1.
 351.9807      213.
 363.9807      147.
 375.9807      275.
 387.9807        9.
 394.9791        1.
 406.9791        1.
 413.9775     3248.
 425.9775      821.
 437.9775        1.
 463.9743     1530.
 475.9743        1.
 487.9743        1.
 501.9711     3660.
 537.9711        1.
 563.9679        1.
 575.9679      143.
 613.9647      300.
```

Figure 114. Listing of ions and expected relative intensities for perfluorotributylamine (PFTBA or FC43) used as a calibration compound in a quadrupole mass spectrometer under EI conditions. (Copyright © 1982, Finnigan Corporation. All rights reserved)

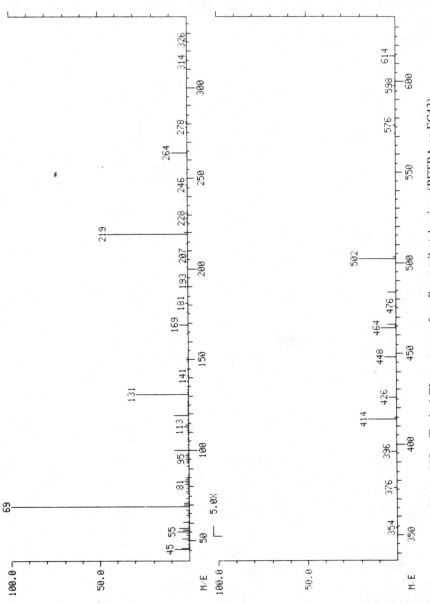

Figure 115. Typical EI spectrum of perfluorotributylamine, (PFTBA or FC43).

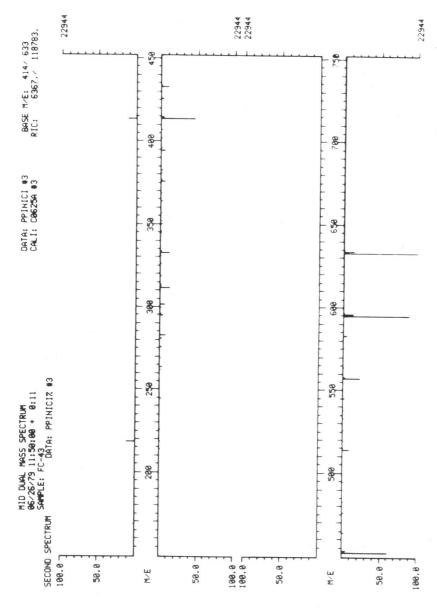

Figure 116. Positive-ion and negative-ion CI mass spectra of perfluorotributylamine (PFTBA or FC43). The negative-ion signal is displayed vertically downwards. (Copyright © 1982. Finnigan Corporation. All rights reserved.)

```
           INPUT FILE:  PFK
          OUTPUT FILE:  PFK
            MASS      INTENSITY
           30.9984        100.
           51.0046        680.
           68.9952      10000.
           80.9952         40.
           84.9655         80.
           92.9952        160.
           99.9936        350.
          113.0014        150.
          118.9920       2500.
          130.9920       2490.
          142.9920        200.
          154.9920        150.
          161.9904        300.
          168.9888       1500.
          180.9888       1490.
          192.9888        300.
          204.9888        200.
          211.9872         50.
          218.9856        700.
          230.9856        690.
          242.9856        600.
          254.9856        300.
          268.9825        400.
          280.9825        600.
          292.9825        400.
          304.9825        150.
          318.9793        200.
          330.9793        400.
          342.9793        300.
          354.9793        100.
          366.9793         50.
          380.9761        400.
          392.9761        200.
          404.9761        150.
          416.9761         50.
          430.9729        300.
          442.9729         50.
          454.9729         50.
          466.9729         50.
          480.9697         50.
          492.9697        100.
          504.9697        100.
          516.9697         50.
          530.9665         50.
          542.9665        100.
          554.9665         50.
          566.9665         50.
          580.9633         50.
          592.9633        100.
          604.9633         50.
          616.9633         50.
          630.9601         50.
          642.9601         50.
          654.9601         50.
          666.9601         50.
          680.9569         50.
          692.9569         50.
          704.9569         50.
          716.9569         50.
          730.9537         50.
          742.9537         50.
          754.9537         50.
          766.9537         50.
          780.9506         50.
          792.9506         50.
          804.9506         50.
```

Figure 117. Partial listing of ions and expected relative intensities for PFK calibration compound under EI conditions in a quadrupole mass spectrometer. (Copyright © 1982, Finnigan Corporation. All rights reserved.)

```
       INPUT FILE:  PFKN
      OUTPUT FILE:  PFKN
         MASS     INTENSITY
        69. 9377      1000.
        71. 9348       700.
        73. 9319       110.
       218. 9856         7.
       230. 9856        37.
       268. 9824         7.
       280. 9824       220.
       292. 9824         7.
       311. 9808        16.
       330. 9792       259.
       342. 9792        20.
       361. 9776        20.
       380. 9760       135.
       392. 9760        40.
       404. 9760         7.
       411. 9744         8.
       430. 9728        50.
       442. 9728        38.
       454. 9728         8.
       480. 9696        17.
       492. 9696        22.
       504. 9696         5.
       530. 9664         5.
       542. 9665         7.
       554. 9664         5.
       580. 9633         2.
       604. 9633         4.
```

Figure 118. Partial listing of ions and expected relative intensities for PFK calibration compound on a quadropole instrument operated in the negative-ion CI (methane) mode. (Copyright © 1982, Finnigan Corporation. All rights reserved.)

tamination is high and the compound should be used sparingly. Figure 120 gives a listing of mass and intensity for perfluoroheptyl triazine run on a typical quadrupole system.

It should be noted that the intensity figures in these lists represent typical reference values for the purpose of a calibration algorithm and are quoted here to give some indication of the relative intensities of ions in each calibration compound spectrum. Intensities and intensity ratios will vary from instrument to instrument.

The heptyltriazine described above is only one of a family of compounds suitable for calibration purposes. For instance, Figure 121 shows the spectra of tris-(perfluorononyl)-s-triazine obtained on an advanced hyperbolic quadrupole system. The molecular ion at 1485 u is clearly visible. As instrument performance and user requirements stretch the working mass range, there will be an increasing need for higher weight reference compounds. These may not be needed for GC/MS work, but often LC/MS interfaces are available to fit the same mass spectrometer system and there is a resultant push for higher mass calibrations.

Higher Mass Reference Compounds.   There is a growing demand among GC/MS users for higher operating mass ranges. As a consequence there is a need for

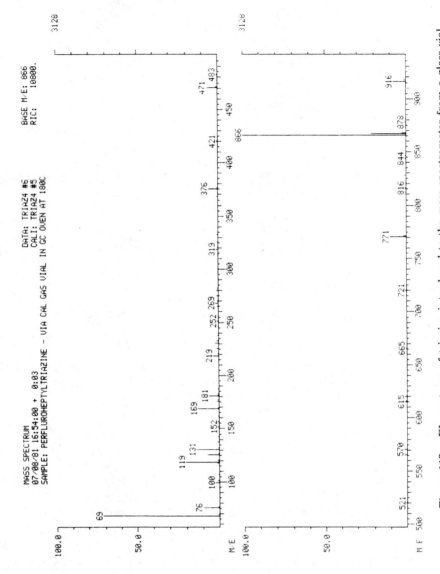

Figure 119. EI spectrum of triazine introduced to the mass spectrometer from a glass vial connected in place of a column in the gas chromatograph. The oven was heated to 180°C. (By kind permission of Finnigan-MAT Ltd.)

```
        INPUT FILE:  TRIAZ
       OUTPUT FILE:  TRIAZ
          MASS     INTENSITY
         19.9812          1.
         27.9949          1.
         39.9624          1.
         43.9898          1.
         49.9968        400.
         68.9952     170752.
         75.9999      10000.
         99.9936       2988.
        118.9920      30944.
        125.9967      10368.
        130.9920      18688.
        137.9967        439.
        168.9888      20096.
        175.9935       1656.
        180.9888      10560.
        218.9856       5328.
        225.9903       1226.
        230.9856       3380.
        268.9824       2360.
        275.9871        505.
        280.9824       1380.
        318.9792       1330.
        368.9760          1.
        375.9807       5696.
        420.9822        318.
        470.9790       5072.
        515.9805          1.
        520.9758        109.
        544.9758          1.
        565.9773        698.
        570.9726        689.
        577.9773          1.
        615.9741        664.
        620.9694        194.
        627.9741          1.
        665.9709          3.
        670.9662          2.
        677.9709          1.
        696.9693          1.
        715.9677          1.
        720.9630          2.
        727.9677          1.
        732.9630          2.
        751.9614          1.
        765.9645          1.
        770.9598       1594.
        777.9645          1.
        789.9645          1.
        815.9613          9.
        827.9613         20.
        846.9597          2.
        865.9581      19136.
        877.9581          2.
        889.9581          1.
        896.9565          1.
        915.9549        832.
        927.9549         94.
        946.9533          1.
        965.9517         91.
        977.9517          3.
       1015.9485          8.
       1027.9485          1.
       1065.9453          1.
       1077.9453          1.
       1089.9453          2.
       1096.9437          1.
       1115.9421          7.
       1127.9421          1.
       1146.9405          1.
       1165.9389         80.
       1184.9373          1.
```

Figure 120. Listing of ions and expected relative intensities for triazine on a quadrupole instrument operated in the EI mode. (Copyright © 1982, Finnigan Corporation. All rights reserved.)

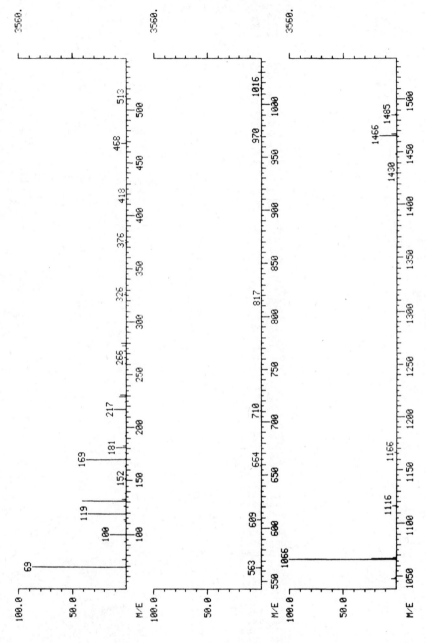

Figure 121. EI spectrum of tris (perfluorononyl)-*s*-triazine obtained with an advanced hyperbolic quadrupole mass filter. (By kind permission of Finnigan-MAT Ltd.)

212

compounds for calibration and tuning purposes which go to higher masses than some of those already mentioned. A number of compounds have been proposed. For instance, Figure 122 shows the spectrum of a mixture of perfluoroalkoxy phosphonitriles. This mixture produces ions at nearly every mass in the range 200 to 950 u, with significant ions above and below this. In fact, the spectrum extends to about 1200 u, 200 mass units above the range of the instrument (a Finnigan 4000 with Incos data system) that was used to obtain this data. Figure 123 gives a mass listing for the more intense ions. The base peak is 887 u followed closely by 919 u. The sample was introduced into the mass spectrometer by spotting it onto the sample belt of the (optional) LC/MS interface that was fitted to the system but could just as easily have been introduced using a solids probe inlet.

Other classes of compounds can also provide high-mass spectra for reference work. Fombolin* is a polyfluorinated polyether that is used as a diffusion pump fluid. With Fombolin in a mass spectrometer ion source, spectra to over 5000 u have been obtained. The regular spacing of ions in the Fombolin spectrum can provide accurate reference points right through and well above the normal GC/MS operating mass range.

SCANNING

The setting of a suitable mass spectrometer scan can be considered as the selection of two variables, mass and time. The mass/time function may be continuous, that is, a smooth transition from one selected mass to the next at a defined rate. This is the normal scan mode for manually tuning the instrument with a real-time display. It is also a common acquisition mode, giving a full spectrum across the defined mass range. Where the sample is known or at least suspected, it is often preferable to monitor particular masses. The time spent at each mass is not necessarily the same. Single masses or small mass ranges may be selected. This type of scanning is known as *multiple ion detection* (MID) or *multiple interval scanning* (MI or MIS) and sometimes as *mass fragmentography*. Single-ion monitoring (SIM) is a particular case of MID scanning. Even though only one mass may be selected, it is still necessary to define a time interval, the acquisition time, if changing signal intensities are to be recorded. Continuous acquisition with no time limit would yield an integrated signal indicative of the total amount of the given mass collected since the start of the acquisition and would probably quickly saturate the signal electronics.

Selection of the total scan time or total interval cycle time is dependent on the chromatographic conditions. The mass spectrometer should be scanned fast enough so that changing sample elution rates during a run do not significantly distort the mass spectrum; 5 to 10 scans across each chromatographic peak are usually sufficient. This sets a total scan time limit of about 3 s for packed column

*Fombolin is a registered trade name of Montediso Co.

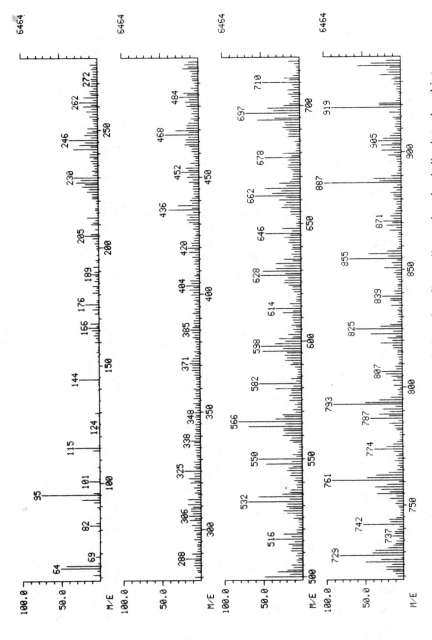

Figure 122. Partial EI spectrum of a mixture of perfluoroalkoxy phosphonitriles introduced into a Finnigan 4000 by spotting onto its LC/MS interface belt. (By kind permission of Finnigan-MAT Ltd.)

MASS	% RA	% RIC	INTEN.	MASS	% RA	% RIC	INTEN.
64	51.30	0.46	3316.	667	26.89	0.24	1738.
65	43.81	0.39	2832.	678	44.74	0.40	2892.
95	75.00	0.67	4848.	694	55.07	0.49	3560.
115	32.12	0.29	2076.	695	31.87	0.28	2060.
144	25.62	0.23	1656.	697	69.68	0.62	4504.
228	28.12	0.25	1818.	698	33.73	0.30	2180.
230	30.91	0.27	1998.	699	40.47	0.36	2616.
242	31.53	0.28	2038.	710	47.40	0.42	3064.
246	37.13	0.33	2400.	726	50.19	0.45	3244.
436	41.21	0.37	2664.	727	29.46	0.26	1904.
438	26.89	0.24	1738.	729	83.91	0.75	5424.
468	43.13	0.38	2788.	730	41.27	0.37	2668.
470	30.69	0.27	1984.	731	36.82	0.33	2380.
484	24.16	0.21	1562.	742	53.09	0.47	3432.
500	49.07	0.44	3172.	755	35.09	0.31	2268.
502	31.93	0.28	2064.	757	29.05	0.26	1878.
516	34.96	0.31	2260.	758	37.75	0.34	2440.
518	24.50	0.22	1584.	761	92.82	0.83	6000.
528	24.66	0.22	1594.	762	46.91	0.42	3032.
532	69.80	0.62	4512.	763	35.71	0.32	2308.
534	55.94	0.50	3616.	774	36.94	0.33	2388.
535	26.14	0.23	1690.	787	41.89	0.37	2708.
548	45.73	0.41	2956.	789	30.35	0.27	1962.
550	55.14	0.49	3564.	791	26.92	0.24	1740.
560	24.85	0.22	1606.	793	90.35	0.80	5840.
564	67.57	0.60	4368.	794	48.70	0.43	3148.
566	80.57	0.72	5208.	795	31.03	0.28	2006.
567	33.60	0.30	2172.	807	27.20	0.24	1758.
569	23.76	0.21	1536.	819	28.59	0.25	1848.
580	33.42	0.30	2160.	823	41.46	0.37	2680.
582	54.46	0.48	3520.	825	60.27	0.54	3896.
596	49.94	0.44	3228.	826	28.68	0.26	1854.
598	51.79	0.46	3348.	839	25.25	0.22	1632.
599	33.85	0.30	2188.	855	67.95	0.60	4392.
601	28.53	0.25	1844.	857	43.44	0.39	2808.
614	32.24	0.29	2084.	887	100.00	0.89	6464.
628	50.99	0.45	3296.	888	27.66	0.25	1788.
630	48.95	0.44	3164.	889	35.27	0.31	2280.
631	32.36	0.29	2092.	901	25.59	0.23	1654.
633	32.49	0.29	2100.	903	25.84	0.23	1670.
646	44.93	0.40	2904.	905	29.98	0.27	1938.
657	23.82	0.21	1540.	919	90.22	0.80	5832.
660	26.83	0.24	1734.	920	27.66	0.25	1788.
662	58.23	0.52	3764.	921	28.37	0.25	1834.
663	35.58	0.32	2300.	933	26.83	0.24	1734.
664	24.10	0.21	1558.	937	55.82	0.50	3608.
665	44.80	0.40	2896.	938	38.43	0.34	2484.

Figure 123. Listing of the major ions in the spectra of Figure 122. (By kind permission of Finnigan-MAT Ltd.)

work and about 1 s for capillary work. Unfortunately, the total scan time cannot always be chosen at will. Most mass spectrometers have an optimum, or at least a maximum, scan speed. If the sample analysis demands a large mass range, for example, with high-molecular-weight compounds, then the scan rate for a given total scan time must be high and may exceed the limits for the instrument. In these cases a compromise must be worked out between the total scan time and the mass range.

Quadrupole instruments can usually cope with high scan rates, but it is wise to allow a significant settling or hold time at the beginning of the scan. This allows the high voltages present at the high-mass part of the scan to decay away so that the low-mass part of the scan is not distorted. A settling time of about 50 ms is usually sufficient. Down-scanning does not eliminate this wasted time. The sudden rise to full mass power at the start of the scan would require just as long a settling time and would put more stress on the driving circuits.

Sector instruments have a variety of scan modes is shown earlier in Figure 33. They are not as fast as quadrupoles but can usually be scanned fast enough for most GC/MS work. If very fast scans are needed, they can be achieved by scanning the accelerating voltage. Accelerating voltage scanning or switching is used extensively for peak matching and MID work. Unfortunately, it is not usually suitable for full scan work or large mass difference MID because of the tuning changes associated with large voltage swings.

Magnets have to be scanned more slowly, and abrupt changes in magnet current should be avoided. Linear current scans, leading to a quadratic mass scale and exponential up scans, are not suitable except for basic real-time displays such as those used for manual tuning. The sudden stop at the end of the scan can cause ringing and instability. It is more common to use a combination of complementary exponential up scans and exponential down scans. These give no sudden changes in magnet current at the top and bottom of the scan. Acquisition, especially with data systems, is made on the down scan because this gives a constant time per mass peak. Continuous scanning under these conditions can lead to very stable and predictable mass scales despite hysteresis effects. Data systems mass calibration can be very precise, even at low resolution. To take full advantage of this it is often better to scan a magnetic instrument continuously over a fairly wide mass range rather than to alter the parameters for each acquisition.

Careful selection of the mass range can usually realize acceptably fast scans despite the limitations of the magnet. Most GC/MS instruments can manage one decade of mass in 3 s and many can go at one decade per second. For low-molecular-weight samples, a single decade, say 20 to 200 u, is sufficient. For higher mass work where scan times must be short, a single decade of 60 to 600 u or 70 to 700 u may be sufficient, providing the possibility of lost low-mass ions is not overlooked. To minimize the total scan time if more than one decade is required, the scan should be started at as high a mass as possible. With all types of instruments it is important to scan over a sufficient mass range so as to overlap the mass range of all compounds likely to be encountered.

So far in our discussion of mass spectrometer scanning we have considered a smooth scan of the mass-set parameter. Under data system control of a full mass scan and in all forms of MID the input parameter is changed in discrete steps. The step sizes are very small for full mass scan work with most data systems, representing a fraction of a mass unit. The system-response factor tends to smooth these out, and they are not significant except at high resolution. Data systems designed for high-resolution work usually have DAC outputs that lead to extremely small mass changes.

Some of the older data systems were designed to step across the mass range from one mass top to the next. One mass unit is a relatively large jump, and a small settling time was usually built in to allow the system to stabilize. MID work requires even larger mass jumps, sometimes of several hundreds of mass units. Settling times should be set long enough to allow the mass spectrometer to reach a steady state. This is usually only a few milliseconds for quadrupole instruments but may be longer for very large mass changes. Sector instruments vary a lot in their MID characteristics, and some experimentation using a known reference compound such as PFK is recommended. As with full scan work, the total cycle time, MID integration times plus settling times, must be fast enough to record peak elution profiles.

It is important to specify mass defect departures from nominal masses with MID work and when using a full-scan data system that jumps from one mass to the next. In some systems this is done automatically, the mass defect value being set for a typical organic compound. Where the compounds of interest are not "typical," and on nonautomatic systems, it is necessary for the operator to select a reasonable mass-defect range. This can lead to unacceptably high scan times if one is not careful. Consider a data system such as the Finnigan 6000 series, which scans the mass spectrometer by stepping from one mass to the next. Mass defects are accommodated by specifying the defect range in tenths of an atomic mass unit. The data system acquires data at each tenth of a mass unit as specified and logs the largest signal as the value to be stored against the nominal mass.

Suppose we wish to scan from 60 to 360 u watching for both positive and negative mass defects. A reasonable mass-defect range may be -0.2 to $+0.3$ u. With a signal integration time of 1 ms and a settling time of 1 ms, the total scan time will be $301 \times 6 \times (1 + 1) = 3612$ ms, or 3.612 s. This may well be too slow for capillary work. At lower masses the mass defect is not likely to be so large, and a smaller range could be specified, for 60 to 210 u, use -0.1 to $+0.1$ u. Higher masses are more likely to give a positive mass defect, for 211 to 360 u, use $+0.1$ to $+0.3$ u, which leads to a total scan time of $(151 + 150) \times 3 \times (1 + 1) = 1806$ ms, or 1.806 s which may well be an acceptable compromise for capillary work. (Note that end masses are included, giving a range of 301 u and not 300 u.)

This type of acquisition is very wasteful of time. Only a twelfth of the first example and a sixth of the second example yields data to be stored. Modern data systems such as Finnigan's Incos system overcome this problem by scanning and acquiring over much narrower mass and time intervals, but the data are summed to give a total mass peak area and are also used to calculate a mass centroid. Mass accuracies to two decimal places are possible even on a low-resolution spectrum scanned at high rates. Mass defects do not have to be specified, because actual masses are measured. Unusual or unexpected mass defects can be identified and can sometimes indicate two ions of the same nominal mass but different accurate masses.

MID is more sensitive than full-scan work because a much longer time is spent acquiring data from the masses of interest. Noise effects tend to cancel out over a longer integration time, so stronger signals with better signal-to-noise ratios are obtained. In the above example, 1 ms was spent on each mass for a reasonable scan time. If only four ions are monitored in the same mass range, about 400 ms could be spent on each, with 10 ms settling time after jumps. This leads to a total cycle time of 1.64 s. Increasing the acquisition time per ion by a factor of 400 may give saturation in the integrator and amplifier electronics. Integration times of 10 to 50 ms are more typical. Cycle times are much shorter and can lead to excessive data rates. Specifying a total cycle time of, say, 500 ms causes an extended settling time at the low-mass ion and reduces the data rate to a reasonable level.

On manual hardware MID systems, high data rates are usually preferred to long cycle-hold times. The integrated signal is usually passed through a filter circuit with a relatively long time constant before being plotted on a strip-chart recorder. The long time constant introduces a delay in the displayed signal, and the recorder cannot follow rapid changes in signal level. Using the higher data rates available, signal-level changes from one integrated sample to the next are small and so the filtering can be relaxed slightly, allowing faster response times with good noise suppression.

RESOLUTION

Another important parameter that must be set by the operator is the mass spectrometer resolution.

Sector Instruments

Sector instruments operate at constant resolution across the mass range, and so the effect of resolution on mass peak width and intensity at both ends of the mass scan must be considered. Setting resolution higher than required at the top end of the scan can result in extremely sharp peaks at the low-mass end. Not only will the peaks be narrow, but also their intensity will be reduced. This represents an overall loss in sensitivity of the GC/MS instrument. However, sensitivity is not the only important parameter from the operator's point of view. Some find that well-resolved spectra are easier to interpret. Higher resolution can be an aid to accurate mass measurement and is necessary to separate ions of different composition at the same nominal mass.

The use of different resolution values can be illustrated by considering the spectra in Figure 124. At low resolution (Fig. 124a), the molecular ion cluster of a brominated hydrocarbon is satisfactorily separated to give nominal mass infor-

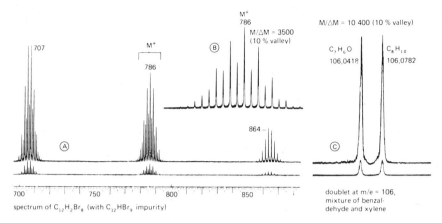

Figure 124. Mass spectral resolution obtained with a MAT 112S sector instrument. (*a*) Part of $C_{12}H_2Br_8$ spectrum (with $C_{12}HBr_q$ impurity) at a resolution of 1300. (*b*) Increased separation in the molecular ion region of spectrum *A* with a resolution of 3500. (*C*) Higher resolution (10,400) is needed to separate this doublet of xylene and benzaldehyde. (By kind permission of Finnigan-MAT Ltd.)

mation for each ion. At lower masses (not shown), ion separation would be greater because the instrument operates at constant resolution, where

$$R = M/\Delta M \qquad (33)$$

Reducing resolution to give stronger ions at low mass would cause the high-mass ion clusters to merge. Resolution must be set sufficiently high to resolve the highest masses scanned, but setting it higher reduces sensitivity. High resolution is no real advantage for nominal mass work but it can improve mass assignments for peak matching or accurate mass data system acquisition. Figure 124*b* shows the same molecular ion cluster at a resolution of 3500. This is insufficient to resolve most ions of the same nominal mass and so will give better mass assignments only if there are no interferences.

Much higher resolution is needed to separate ions of the same nominal mass. Figure 124*c* shows two ions at mass 106 u. Their separation is 36.4 millimass units and requires a minimum resolution of about 3000 just to separate them. In the example shown, a resolution of 10,400 is necessary to give a good baseline separation. To obtain the same degree of separation for the same mass difference between ions at a nominal mass of 510 u a resolution of 50,000 would be required. Instruments that can give such high resolution must be scanned slowly, too slowly in fact for most GC/MS work. Very high resolution work tends to be limited to the analysis of samples introduced over a significant time period. The resolution limit for most GC/MS systems when using capillary columns is about 3000 because of the scan speed limitations. Operation at higher resolution reduces scan speed so that resolutions of up to 10,000 are effective only when using packed columns.

Apart from the mass spectrometer scan-speed/resolution constraints, the data system conversion rate must also be considered on automated systems. The MAT 200 spectrosystem is a very fast data system for high-resolution work, but even so it is limited to 0.4 s/decade at a resolution of 500 and 4 s/decade at a resolution of 10,000 if instrument performance is not to be impaired. In general, when doing GC/MS work it is better to let the gas chromatograph do the separation and the mass spectrometer the identification.

Quadrupole Instruments

Quadrupole instruments operate with a constant mass peak width and so resolution, as defined in Eq. 33, is a function of mass. In fact, the resolution for a quadrupole mass spectrometer is often quoted as

$$R = Am \quad \text{up to mass } M \tag{34}$$

where m is the mass of interest and A is some constant, usually 2, 3, or 4. M is the upper mass limit of the instrument. Low-mass ions, which tend to be more intense anyway, are enhanced by this mass-dependent resolution feature compared with high-mass ions. Comparing quadrupole and magnetic instrument spectra gives the initial impression of poor high-mass performance on the quadrupole instrument. In fact, most manufacturers quote similar sensitivity specifications for the molecular ion of methyl stearate in terms of signal-to-noise ratios. However, when two spectra are compared, the magnetic instrument will always appear to give better high-mass performance because the spectra are usually normalized to the base (lower mass) peak. Typically, the methyl stearate molecular ion at 298 u is 20% of the base peak of 74 u for magnetic spectra, whereas 5% is normal on a quadrupole. If the two spectra were normalized to the molecular ion, the quadrupole would appear about four times more sensitive at the base peak. This implies that quadrupoles should give greater total ion-current sensitivity, especially for compounds that show a lot of fragmentation.

In practice, things are not quite so straightforward. Quadrupole systems generally have higher noise levels, and although total ion signals may be greater, so too is the noise on the RIC trace. It would not be wise, therefore, to make bold statements about the relative sensitivity performance of magnetic and quadrupole instruments. On all instruments, overall sensitivity increases as the resolution is reduced. This means that the operator of a sector instrument should select the lowest practical value of resolution consistent with the likely mass range of the sample to obtain the best sensitivity. The operator of the quadrupole instrument is not faced with this decision, as the resolution alters automatically across the mass range, but he or she should choose a mass peak width that will give the required degree of adjacent mass separation. This is especially important when doing quantitative work.

Having dismissed high-mass discrimination in quadrupole instruments, one should be aware that falloff can occur in high-mass performance. The quadrupole rods are aligned to very fine tolerances, and anything that affects this alignment either mechanically or electronically will have a marked effect on high-mass performance. The rods should be handled carefully and kept clean. For high-mass work it may be necessary to operate at ion energy and multiplier voltages higher than normal—a higher ion energy voltage to get the ions through the quadrupole, and a higher multiplier voltage to get them round the x-ray shield. Resultant signal intensities at low mass can be very high.

Using ion-energy programming, even operating so that the ion energy at low mass is lower than optimum, helps to prevent saturation of the low-mass ions while giving higher values at the top of the mass scale. The use of conversion dynodes, not just for switched positive/negative CI work but also on EI instruments, overcomes many of the problems caused by having to operate at above-normal multiplier voltages, and manufacturers are now fitting them as standard equipment on some systems.

Although resolution varies automatically with mass on quadrupole instruments, the operator can usually set the basic mass peak width; that is, the value of the constant A in Eq. 34 can be selected. One generally likes to operate with as sensitive an instrument as possible, and so mass peak width is usually adjusted to give only baseline separation or even 5 to 10% valleys. Mass peak width should be reduced to raise resolution for quantitative work. This prevents an intense ion from contributing to isotope or adjacent mass intensities. Good baseline resolution helps with data systems mass assignment and identification of doubly charged ions at half-masses. Some instruments have a number of preset resolution settings, and it is normal to set one for ultimate sensitivity, nominal mass work and another for accurate mass, half-mass, or quantitative work.

DATA ACQUISITION WITH A MANUAL SYSTEM

Once the mass spectrometer is set up, data can be acquired. This may be in the form of a strip-chart recording on manual systems and stored and displayed data on automated instruments. These two approaches to data acquisition have some common features, but in general the techniques are quite different. Nevertheless, those users with automated instruments sometimes have to make a manual recording or at least run a sample without the benefit of data system control and data capture. In some cases failure of all or part of the data system shuts the instrument down, but on many instruments it is possible to run a limited service. A familiarity with manual techniques will be useful if ever the data system is out of action.

On manual systems, chromatographic data are usually recorded as a strip-chart trace of total ion current, TIC, while spectra are recorded, when required by the operator, on UV light-sensitive paper, using a light beam oscillograph. TIC data can be obtained on sector instruments from a total-ionization monitor, TIM, a plate that intercepts a small but fixed percentage of the ion beam. The output signal is proportional to the total ion signal leaving the source and is not dependent on the scan position of the magnet.

On quadrupole instruments, it is not possible to use a total-ion monitor, and so the TIC signal is reconstructed by integrating the mass spectral signal from the multiplier and amplifier circuits. This technique can also be used on sector instruments. The actual signal obtained is scan dependent and will not always give a true picture of the GC performance, especially for very sharp GC peaks that are scanned slowly. On the other hand, the RIC trace can be obtained from a portion of the scanned spectrum. The integrators can be set to start at some mass other than the first mass in the scan. This has the effect of suppressing significant low-mass background ions and can give a chromatogram with a more stable baseline. Care should be taken to ensure that the Integrator Start position is set at a mass of low intensity or in the valley between two strong ions, or it will introduce a noise signal due to integrator start errors. A similar argument also applies to the Stop Integration mass, which is usually the end of the scan. The scan should be adjusted so that it does not end at a strong background or bleed ion.

Ultraviolet Recordings of Spectra

Ultraviolet recordings may be made with the mass spectrometer scanning at the speed set for the whole of the GC/MS run, but slower recording scan speeds are often preferable. Typically a quadrupole instrument will scan the selected mass range 15 to 20 times a second to produce a flicker-free real-time oscilloscope display as well as an integrated RIC output. This gives a scan rate in excess of 5000 u/s for typical mass ranges and is too fast for normal UV recording of spectra. Recorder select, speed, and filtering controls usually provided so that the operator can start a slower scan, 100 u/s being a typical speed. This gives a total scan time in our example of about 3 s and will cause a change in the strip-chart recorder RIC output.

On most instruments the signal integrator is disabled during the recording scan, the output being held at its last value. This produces shoulders or steps in the RIC trace and distorts the chromatogram. If an undistorted trace is required, the sample should be run twice on the GC/MS system, the first time without recording spectra. Alternatively, the sample should be run under similar circumstances using the same column if possible, first into a standard gas chromatography detector and then again into the mass spectrometer. One can not only obtain undistorted traces by these techniques, but also anticipate the optimum points to start the UV recording scan so as to take spectra near the tops of the chromatographic peak.

Magnetic instruments are usually scanned more slowly, 1 to 3 s per decade is usual, leading to total scan times in the range of 2 to 5 s for most GC/MS work. These speeds are compatible with UV recording of spectra, and a recorder speed setting is not usually required. Even so, the feature may be provided, especially if the instrument is capable of operating at medium to high resolution, for which scan speeds usually have to be reduced. On very-high-resolution instruments, scan speeds are usually so slow that spectra can be recorded with a strip-chart recorder instead of a UV recorder.

Increasing the number of GC peaks and/or decreasing their peak width makes it more difficult for the operator to select the optimum starting points for a UV recorder scan. Such conditions are common in capillary GC/MS work. One solution to the problem is to leave the UV sensitive paper running continuously. It can then be cut up into consecutive spectra for examination after the run. This, of course, uses a lot of expansive UV-sensitive paper and may disable the RIC output. Relating spectra to chromatographic peaks is then a problem, unless the run can be accurately reproduced. Often only certain parts of the trace are of interest, and a compromise can be made by reducing the scan speed and switching on the recorder a little before the areas of interest.

The problem of the lost RIC trace is not so easily overcome and depends on the instrument manufacturer. A slow-scan RIC output is not usually provided because of the dangers of integrator saturation and leakage losses over the relatively long acquisition time. Even where RIC outputs are available at low scan speeds, one should be aware of these problems and not rely on the RIC trace for accurate quantitative work.

Hardware MID Acquisition

The data-acquisition arrangement is changed somewhat when selective ion monitoring is used. The hardware control unit usually contains the necessary electronics to integrate the individual ion signals and output them to a multipen recorder. Apart from the recorder output signal gain (relative sensitivity), the operator can usually choose the integration time for each ion. This time should not be set too long or there may be a loss of signal due to leakage effects. In the full-scan example given for a quadrupole system, the integrator had to acquire and hold a scan of data for about 60 ms. A similar maximum time is reasonable for MID work. With full scanning, much of the time is spent between mass peaks. MID may be used to acquire data continuously from a mass peak top, so integrated signals may be much greater. A typical integration time is 10 ms, but a range of values from 1 to 100 ms is often provided.

One must also consider the background level when deciding on the integration time. This, too, will be integrated and can give a large output signal. Most hardware MID units have offset controls that reduce the effect of steady background signals. Even so, on should try to choose ions or conditions that do not give strong background signals at the masses to be monitored. Pressure fluctuations and background level changes with column temperature increase the noise and drift on the signal baseline, limiting the level of detection. Column bleed, for instance, will often increase as a compound elutes, being carried out, as it were, with the sample.

Partial-pressure changes in the source as compounds elute can often reduce background signals momentarily and give rise to significant steps in the MID trace that cannot be filtered out. If possible it is best to monitor ions where there is no interfering background regardless of the fact that a hardware zero bucking control is available. One should then switch off or set to zero any output level control and zero the recorder appropriately. This prevents a loss of dynamic range at the recorder due to saturated output amplifiers at the MID unit and on the recorder channel input.

Unlike full-scan work, MID involves jumping from one mass to another in a nonuniform way. The mass spectrometer signal should be allowed to stabilize before a measurement is taken. The necessary settling time should allow for two main elements. First, it should be long enough for the mass spectrometer to settle at the new mass and for the ion transmission signal to stabilize. Second, if the signal level change is large, the input preamplifier may take some time to settle at the new signal level. This time may be 10 or 20 ms for high-gain settings. The situation is further complicated if the preamplifier gain is changed for each ion. However, most MID hardware units have preset settling times dependent on the control settings, and the operator need not be aware that they exist except that the settling times have a bearing on total cycle time. Consider the timing sequence for the three-channel quadrupole MID unit shown in Figure 125. Channels 1 and 3 are set with a moderate gain. The mass set and gain settling is relatively quick at 2 ms, and the integration time for each of the two channels is set at 10 ms. Channel 2 is set for a very low-level signal. The gain setting is high, and so the mass set and gain settling time is increased to 6 ms. The integration time is also longer at 100 ms. This gives a total scan time of 130 ms.

Where there is a possibility that the high-gain, long-integration-time channel may be exposed to a strong signal (channel 2 in this case) there is a danger that there may be leakage, breakthrough, and charging in various parts of the circuitry, giving a memory effect during the next channel period. Such a situation is possible in residue work when a standard sample is run or if a gross residue is detected. This is also a possibility for the other channels but is most likely to occur with saturated or nearly saturated high-gain circuits. The problem can be overcome on systems with fixed settling time by operating with dummy low-gain channels just after the high-gain channel. The mass may be set either to the mass of the channel following or at some suitable low-level ion in between the two important ions. In our example, adding a fourth channel in this way with an integration time of 10 ms and a settling time of 2 ms increase the total cycle time to 142 ms.

To obtain reasonable chromatographic peaks or mass chromatograms in the case of MID, one needs 10 to 15 data points across the peak. This sets a lower limit for the chromatographic peak width of 1.3 s in the former case and 1.42 s in the example with the dummy channel 3. This is satisfactory for most GC/MS

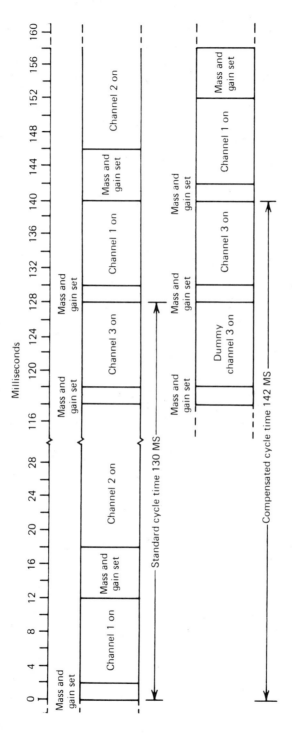

Figure 125. Typical timing sequence for a hardware MID unit showing standard timing and a method of providing extra settling time.

work. Increasing the number of ions monitored, the integration times, or the settling time (magnetic MID units tend to have longer settling times) will cause the total cycle time to rise and the minimum measurable peak width to increase. For example, on a four-channel MID system with each channel set for an integration time of 100 ms and a settling time of 6 ms, the total cycle time would be 424 ms and minimum chromatographic peak width would be 4.24 s. This increases to 4.72 s if dummy channels are used. This is fast enough for packed column work but may be marginal for an accurate capillary response. In this circumstance, reducing the number of ions monitored or the integration time of some of them will speed things up sufficiently for a reasonable capillary response.

Even with packed column work it is undesirable to operate with too many channels with long integration times. In between integration times on a given channel the output is held at its last level. During the channel setup time, the integrator is cleared for a new sample. At the end of the integration time, the integrator signal is passed to the hold circuit. The output of the hold circuit may show large changes in level, especially if the cycle time is long. This tends to distort the mass chromatogram trace. Filtering at the output reduces this effect, but as the time constants need to be large, peak slewing, tailing, and reduction of height may result. Quantitative data may then be suspect. The answer once again is to reduce either integration times or the number of channels to give more data points per chromatographic peak.

DATA ACQUISITION WITH AN AUTOMATED SYSTEM

As already discussed, acquisition with a large number of compounds and/or narrow chromatography peaks leads to difficulties with data capture in terms of UV recording of spectra. In additional, there is no second chance to go back and look at small peaks or background signals. Recorded spectra contain background ions as well as sample ions and must be encoded manually, a technical and time-consuming job. Data capture and handling with a computerized system overcomes these problems and opens up several additional possibilities. For instance, plotting the response for just one characteristic mass after acquisition can identify scans of interest for further study. Mass assignments, spectral labeling, background subtraction, and record keeping can all be done automatically, not to mention all the calculating power available in terms of computing once the data system has been added. All this is not without its constraints, and a number of additional points must be considered when a data system is used.

Zeroing

It is vital that the data system be correctly zeroed. Data acquisition involves the digitizing of the acquired signal, but the ADC system is usually configured to accept signals of only a single polarity. That is to say, signals of only one polarity are satisfactorily converted into a digital signal. As the signal level approaches zero the digital output also approaches zero. Unfortunately, amplifier circuits can have dc offsets so that they may give zero volts output even though there is a small signal present from the mass spectrometer. As this ion current reduces to zero, an analog output will swing through zero to the opposite polarity. A mechanical or physical zero on the display unit (oscilloscope or recorder) can be established even though this is not electrically at 0 V. Data system ADCs are not usually configured to cope with this, and once the 0 V level is reached, the output will stay at zero, even though the signal polarity reverses. As a result, small signals will be lost. Figure 126 illustrates the effect of zero settings that are too high or too low on detection level and isotope ratios, which can both be drastically affected by poor zeroing.

Another problem associated with zeroing on GC/MS data systems is the identification of the start and end of a mass peak. If the signal does not come down to the baseline, the data system cannot be sure that one mass peak has finished and another started. Most data systems that acquire data by scanning uniformly across the spectrum have an algorithm to resolve peaks provided there is a reasonable valley between them. Problems occur when the valley is not well defined, due to noise, low signal levels, or half-mass, doubly charged ions, and these problems are exaggerated by poor zeroing. Quadrupole instruments operating with a mass-dependent resolution suffer in this respect more than magnetic instruments. Even so, definition of mass centroids is complicated by a lack of zero-volts resolution even though there may be baseline separation, the baseline being "zeroed" above 0 V.

Figure 127 illustrates this problem and also suggests a method of overcoming it. The reconstructed mass chromatograms for three adjacent masses are shown. These are the signals that were stored for each mass as part of a wider scan. A fairly low-level ion was chosen as the middle mass. In this case, 264 u from a very low bleed rate of FC43 calibration compound was used, but any steady low-level ion signal is suitable. The quadrupole mass spectrometer peak width was defined as 1000 mmu in the data system parameters. For scans 1 to 75 the zero was set too high, and the noise level at mass 264 u (center trace) was significant. This noise is in fact mass assignment dropout due to zero error and can be seen as noisey signals at masses 263 and 265 u. The acquisition direction was up, so the noise signal level is greater at 265 u than at 263 u. This may not be obvious at first glance because of the independent display normalization. The vertical axes are scaled at the left-hand side in percentage terms and at the right-hand side of ADC counts.

The zero was turned down at scan 75 and again at scan 100 and the noise levels were reduced dramatically. In the region of scans 100 to 130 the zero was probably set too low, and between scans 140 and 150 it was adjusted up slightly to an acceptable level. The occasional ion was now detected at masses 263 and 265 u with a steady, low noise signal at 264 u. The adjustments were made while

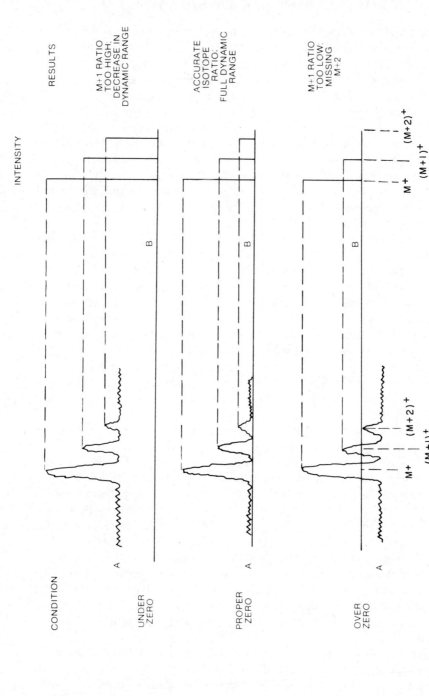

Figure 126. Effect of analog zero on digitized spectra. (Copyright © 1982, Finnigan corporation. All rights reserved.)

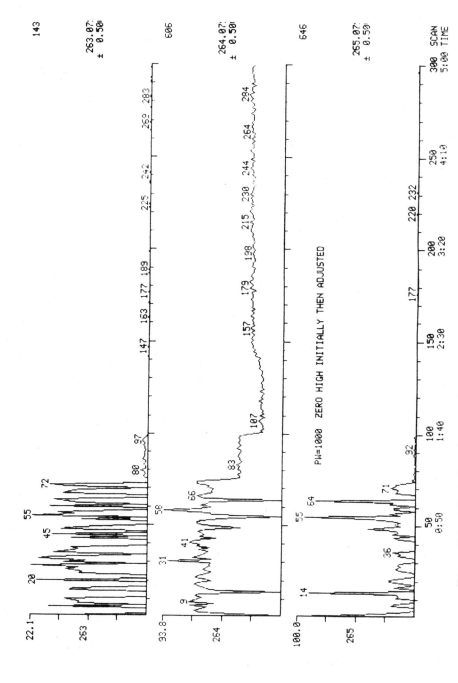

Figure 127. Effect of zeroing on noise and mass assignment. The initially high zero was adjusted and optimized between scans 70 and 150.

the instrument was acquiring data with the mass chromatograms displayed in real time. Adjustment of the zero so that a low signal level could just be observed gave satisfactory results in terms of both isotope ratios and data rates. One consequence of too high a zero setting is the amount of "signal" data that is needlessly acquired. Storage space can soon be filled with meaningless data of small zero-error peaks at every mass.

Most GC/MS systems have several zero adjustment points to set the levels of the various amplifier circuits. It is recommended that the whole system be properly zeroed before trimming for the data system zero. This has the advantage for both manual and data system work in that the signal baseline does not change much when gain and sensitivity controls are altered. Displays and traces stay on scale, and the fine data system zeroing is then just a minor adjustment to the main operator zero control.

Data systems that step from one mass to the next in discrete jumps do not suffer from the problem of mass dropout due to baseline zero errors. Even so, proper zeroing is vital for satisfactory signal detection and isotope ratio assignment. The foregoing procedure outlined cannot be used to zero the instrument, and references should be made to the system operating manuals for the correct zero adjustment method.

Integration Times

On some systems, integration time must be defined by the operator, and the resultant scan time is the sum of all the integration times for the masses selected plus the defined settling times. Slower scan times can be selected and a delay is added to the beginning of the scan to achieve this. On other systems, scan time and the number of data points per mass peak are defined and the necessary integration time is calculated. A variation in this approach that is used in most modern data systems involves an analog-to digital convertor (ADC) preceded by an integrator with a fixed integration time. This generates an ADC sample of fixed interval. A number of these ADC samples are then added, depending on the scan conditions, to give a centroid sample. The centroid samples may be stored directly, giving details of peak shape (profile data), or used to calculate mass centroid and total mass intensity (centroid data).

Whichever system is used, the operator has some control within the limits of the other instrument parameters. Where adjustable or selectable, the integration time should be neither too short nor too long, 1 to 5 ms being typical. Short integration times lead to excessive noise, whereas longer times risk saturation of the integrator or at least a high signal level. At high levels and long integration times, there is a greater likelihood of electrical leakage, giving distorted peak heights. On systems with fixed integration periods, enough summed samples (centroid samples) should be requested so that the center of mass can be calculated sufficiently accurately. Ten to fifteen samples per mass peak are typical, with higher numbers for higher mass accuracy.

Noise Rejection

Analog (manual) systems are usually fitted with a filter, defined either in terms of bandwidth or in atomic mass units per second. This can reduce noise on the acquired data. Analog filtering is not usually applied to the signals when a data system is used; the raw data are fed straight to the computer interface. At this point the designer of the interface has a number of options that can significantly reduce both noise and data rates. Integrators and the summation of integrated samples act as noise filters up to a point. Software constraints can then be applied to validate the data. The most commonly used parameter is the threshold level. Setting a threshold level to discriminate against low signal and noise levels seems similar to setting the zero high. The difference is that if the threshold is exceeded, then all the data, and not just those above threshold level, are passed through.

Further discriminating software filters such as minimum width and area may be applied. Depending on the system design, the minimum width is the minimum number of ADC or centroid samples that must exceed threshold before they are saved as valid data. Threshold and width are useful noise reduction parameters for low-level signals. *Area* refers to the total integrated area of a mass peak once it has been defined. Setting a minimum area is useful in reducing data rates, since only peaks of a significant size are saved.

The action of these various filters is illustrated in Figure 128. Most of the baseline noise is below threshold and is rejected. Spikes that exceed threshold are present for only a short time and will not exceed the width filter. Mass peak 3 may exceed the width criterion and can be accepted or rejected on the basis of its area. Mass peaks 1 and 2 exceed all the filter and area criteria and are accepted. Unfortunately, they are poorly resolved. They may be merged into one peak or separated into two, depending on the fragmentation software.

Accurate Mass Acquisition

It is possible with some data systems to acquire data as a series of points across a mass peak, which are then used to calculate the peak centroid and intensity. The data are usually displayed or at least available for display soon after they are acquired, so the mass calculations must be relatively rapid. More accurate mass assignments can be made using slower algorithms and by referencing to other masses in the spectrum. The operator often has the option of saving the data in basic centroid mass and intensity form or as the raw peak profile data for later mass calculations. Obviously, data storage is used up faster when the profile mode is used.

Another way of improving mass accuracy with sector instruments is to operate at higher resolution with a reference compound bleeding into the source at the same time as the sample. The data system then gives mass assignments based on mass ratios to ions in the reference compound. This technique is valid only if the

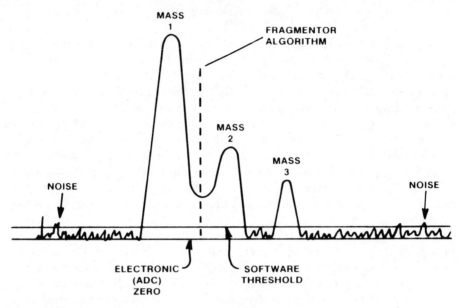

Figure 128. Peak definition and noise rejection in digital acquisition systems. (Copyright © 1982, Finnigan corporation. All rights reserved.)

resolution is high enough to resolve satisfactorily the sample and reference mass peaks and if the scan speed is relatively slow and stable. Often it needs to be slow for satisfactory resolution performance of the mass spectrometer. Also, one must allow for the large number of mass peaks when the calibration compound is added to the sample. Fast scan speeds may give unacceptably high data rates.

Digital control of the scan may also be a problem with high-resolution data system acquisition, the mass-set control causing peak jitter, which appears as signal noise. For very high resolution work it is normal to set the mass spectrometer scanning in a manual mode and acquire the data on the fly. Mass assignment is difficult, so it is usual to operate in this mode with a calibration compound such as PFK continually bleeding into the source. Care should be taken to set PFK level as low as practical so as to minimize contamination and maximize source operating lifetime.

Data Rates

Early GC/MS data systems were acquire-only or process-only systems. Modern systems can acquire data from one, two, or more GC/MS instruments or other inputs and still allow data processing. The various jobs are usually time-shared but happen so fast that they appear to be simultaneous. As the work load increases there is usually an obvious reduction in the time available for data processing. The operator-related jobs, such as data display and manipulation are said to be "background jobs," while GC/MS control and data acquisition, which have a higher data system priority, are called "foreground jobs." As the foreground jobs increase, with high data rates for example, the time available for background work is reduced. It is therefore important to optimize the data rates during GC/MS acquisition by setting reasonable acquisition filters, getting the zero properly set, and minimizing the sample size. Not only does this make the best use of the available storage, but is also maximizes the background operating time during acquisition. A similar effect occurs when more than one data terminal is connected to the system. Each terminal calls upon the computer for attention, and there may be noticeable delays as the computer switches between them. These delays become even more pronounced during acquisition, so it is very important to optimize data rates on dual-terminal systems.

Of course, not all data systems suffer from these problems. Some, usually older, systems cannot be used for processing during data acquisition and the problem does not occur. Others have multiple processors and can handle much higher data rates. Each function or unit is handled by its own processor, and all the processors operate as slaves to a main controller. The arrangement is known as *distributed processing*, and each processing element provides localized computing and initial data storage. Data transfer to and from the main control system can then be more efficient in terms of access timing. There are more computing elements in a distributed processing system, and they are just starting to find application in GC/MS data systems as the relative unit cost of processing and storage elements falls to economic levels.

DATA HANDLING

No discussion of data system acquisition would be complete without mention of the subsequent data handling. This is an area that has seen tremendous growth in recent years, and the pace of development continues. Detailed discussion of the wide variety of software routines available from each system manufacturer is outside the scope of this work, but some general comments are appropriate.

Software

Data systems are only as good as their programs and the raw data supplied by the GC/MS instrument. In most cases the software routines will be tried and tested (debugged), and programming faults are rare. However, there is always more than one way of doing things, and many routines could be improved or at least given better features. On some systems operators can try their hand at programming or at least at writing procedures that link available software. To avoid much heartache, especially with long and complicated programs, it is recommended that original master software not be edited. If it seems desirable to modify a program or procedure, the original software should be copied so that the system can easily be reinstated if the changes are not successful. Often this can be done at the program file level rather that at the whole disk level by renaming files or copying to new file names. It is essential to document changes so that a

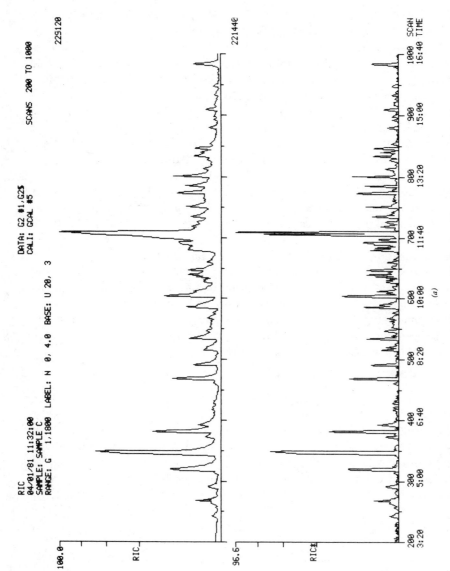

Figure 129. (*a*) Data enhancement on a long capillary run. Top trace, row data; lower trace, enhanced data. The program appears to have separated a peak at scan 710 into two compounds.

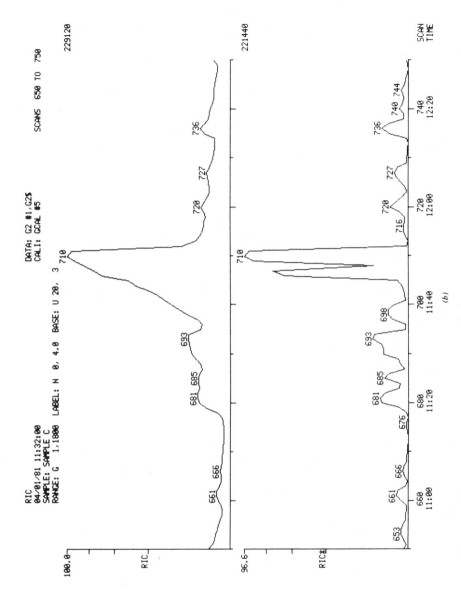

Figure 129 (*continued*) (*b*) An expanded portion of the data shown in (*a*). Although not obvious here, the top, raw, data were based on spectra with saturated peaks.

235

reversal can be made or correct "old" files can be deleted when the new program or routine is successful. It is also important that the modifications be adequately documented so that they can be understood by others (and by the originator) if they have to be inspected or further modified at a later date. This documentation can often by included within the altered program in the form of comment statements.

Sometimes improvements and changes may seem desirable but the necessary programming is beyond the operator or impossible to achieve on the standard operational system. Manufacturers always welcome constructive criticism of their products, and all have a good record of implementing customers' ideas. This is especially true in the area of data systems software. Data system users are therefore strongly urged to communicate their ideas to their system's manufacturer so that users generally may benefit from the system development.

Quality of Data

Computerized systems also depend on reliable data to produce satisfactory results. This can best be summed up by the old quip "Garbage in, gargage out!" The data system cannot improve poor data, and the onus is on the operator to achieve satisfactory GC/MS results. On the other hand, data systems are very good at adding credibility to otherwise suspect results. Just because the system produces nice bar graphs and numerical results to several decimal places, one should not jump to the conclusion that output is valid and accurate. Results should be checked—not recalculated by hand, which in many cases would be impossible, but considered for validity. The data system should not be used to mask poor GC/MS practice or to produce arbitrarily more accurate data.

Having been cautioned against overoptimism on its capabilities, the data system user should not lose sight of the fact that a tremendous amount of information can be obtained. Much is useful and much is not. One should give some consideration to the use of the data system and the required data presentation rather than just using the facilities because they are there. A couple of examples best illustrate these general points.

Most data systems have a library search facility with a resident reference library such as the NIH/NBS library. At the time of writing, the latter has 31,000 entries, which may seem a large number. It is therefore tempting to think that all compounds that are likely to be encountered will be in the library. When one considers the possible combinations of even a small number of elements that can form an organic molecule, one quickly realizes that the library is far from complete. In view of the millions of molecules entering the mass spectrometer, 31,000 library entries is a relatively small number.

Library search results should be considered only as indicators that warrant careful inspection before the answers presented by the system are accepted. Often the class of chemical compound is identified but the library fit results are poor. It may be that the compound is simply not in the library. Do not discount the results completely if the matches seem low. The data system picked them for a reason, and that reason warrants investigation. Used in this way, the library

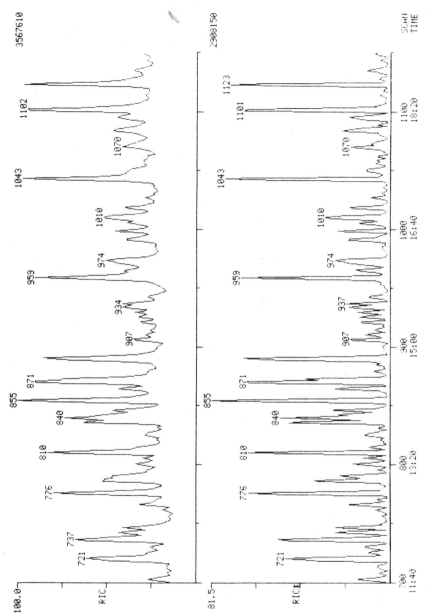

Figure 130. Part of a chromatogram of a diesel oil spill sample. The top trace is the raw GC/MS data. The lower trace is the same data after a computer enhancement routine. Note the improved separation of peaks around scan 840 and between scans 950 and 980. The enhanced data is pulled to the baseline, which reduces data storage requirements and improves presentation. (Courtesy of Robin Law)

237

search routines are an aid to better GC/MS analysis and not just a replacement for the skilled operator.

Another feature often available is data enhancement (Fig. 129a). The many different ways that this can be achieved and their various merits are not under discussion here. Suffice it to say that their aim is to deconvolute unresolved or poorly resolved GC/MS peaks by considering the changing mass spectral patterns. Figure 129b shows an expanded portion of a long capillary GC/MS run, possibly with unresolved peaks in the range of scans 700 to 720. This is apparently confirmed by the enhanced trace shown below the raw data; two peaks are shown. Closer examination of the data reveals that the gas chromatograph was overloaded and the mass spectrometer output was saturated for several ions in these scans. The enhancement routine was in error, because it was working with poor-quality data and, as a consequence, produced an anomalous result.

On the other hand, one should not write off enhancement routines with a few strokes of the pen. In many cases they can pull the chromatography peaks out of the background or separate two overlapping compounds. Another feature of enhance routines is that a great deal of unwanted, irrelevant data are discarded. The storage requirements for these data may be greater than for the data of interest, and so enhance routines can be very beneficial in minimizing data storage requirements. Figure 130 is typical of the improvements that can be obtained.

Unfortunately, a word of caution is once more necessary. Enhancement routines often change spectral peak ratios, especially when background ions are present at significant masses. This is rarely a problem except when quantitative analysis is required. Always check quantitation results on enhanced data carefully before discarding the original raw data. On some systems it is possible to enhance limited portions of the GC/MS run, leaving critical peaks as raw data. In these circumstances the operator has the best of both worlds.

Data Presentation

The user of computer displays, an integral part of modern data systems, has given us a wide range of data-presentation formats. Some are illustrated in this text. The rate of development in this area is considerable, and any attempt to review the topic will soon be out of date. The traditional terminal is the Tektronix type with a green phospher, but some companies, including Tektronix, are offering alternatives including multicolor presentations.

Most users find that they settle down to a few standard formats and soon learn the relevant interactive commands. The experienced user may find it worthwhile to reread the manuals and experiment with new procedures to broaden his or her knowledge of the system. Often the system has the capability to present and organize data in subtle ways that were not seen or understood when the user was new to the system. This is especially true in terms of hard copy output. A lot of report work as well as the basic data presentation can be achieved with the computer.

CHAPTER NINE

Laboratory Practice

There are several approaches to running an analytical facility, and the success or failure of a particular approach will depend in the main on the individuals involved. The fact that a laboratory always looks neat and tidy does not guarantee success. A scruffy lab can produce very high quality data; it all depends on the commitment of the instrument operators and their attention to critical points of detail. However, there are a number of points that favor the clean and tidy approach.

Generally people will not admit to contributing to the mess; it is usually someone else's fault. Everyone would like to work in a clean and pleasant environment in a well-organized fashion if only they did not have to make the effort because nobody else will. In this situation it is the system as defined by those responsible—the managers and supervisors—who let the side down. Responsibilities must be assigned and standards established so that everyone in the laboratory knows what is expected with regard to care and use of the equipment. Assigned responsibilities and defined standards then need to be monitored and reviewed as part of the general review of each individual's work.

Rules and regulations are not sufficient; there must be motivation. Apart from the direct benefits associated with working in a good environment, there is the satisfaction derived from positive judgments expressed by others. In closely associated groups, actual results will count for more than anything else, but as the relationships become weaker or more remote, appearances will play an increasing part. This is true in all walks of life, and one should not assume that results speak for themselves. Laboratories have lost vital funding because their general appearance has reduced their scientific creditability.

Quite apart from the esthetics of "running a tight ship," there are increasing legal requirements in various industries for documented procedures, protocols, and practices. Since one is being forced to do these things it makes sense to apply them to achieve the maximum return. A major argument against this philosophy in general and regulations in particular is that the documentation and record keeping divert the scientist and technician from the real work at hand and thus reduce operating efficiency. In my experience, the opposite is true. In orderly laboratories that "follow the book," procedures become routine and, once established, create no extra work. There are direct benefits to be gained from information that is stored, because it is accessible to all. In laboratories where organized chaos is the order of the day, everything depends upon the memories of individuals. The situation is very intolerant of absence due to sickness, holidays, or even brief meetings elsewhere. The loss of information when people leave the organization can be catastrophic.

A clean laboratory, attention to procedures, and detailed records may not always produce satisfactory results, but they must play their part in the running of a GC/MS facility. If used to advantage, they can increase operational efficiency.

RECORD KEEPING

It is essential to keep records of the principal operating parameters of each GC/ MS run so that the analysis can be repeated or compared directly with similar work. On manual systems the records must be handwritten on the UV and strip-chart recordings. All too often, part or all of the record is left out. To simplify matters a standard form could be produced either as a ticket that is stapled to the traces, such as the gas chromatography version shown in Figure 131, or as a printed sheet that is filed separately but with information included to locate the appropriate traces (Fig. 132).

Another approach is to keep a log book for all the parameters. If the pages of the book are prepared with columns for the various details, then the operator need only write in the numbers. This is illustrated in Figure 133. In fact, only those parameters that change need be recorded. In this way a lot of information can be saved, and trends in the instrument's tuning can be monitored by scanning down a column. Over a period of time it is possible to establish rates of change with numbers and types of some samples and so predict when source cleaning is necessary. Detailed record keeping of this type is useful with a new instrument as its normal pattern of operating parameters is established.

Service and maintenance information is probably best kept separately. A diary is quite useful for this, and an entry should be made each day by each user with brief comments about general performance. Any serviceability problems or maintenance tasks should be also recorded along with details of service visits. Results of standard tests and due dates for tests, filter changes, and so on can also be recorded. In this way a serviceability picture for the instrument can be established and routine maintenance planned. The diary also provides a record of users in a multi-user situation.

Users of automated instruments can often make use of standard log files available on their data systems to record relevant information. As with manual

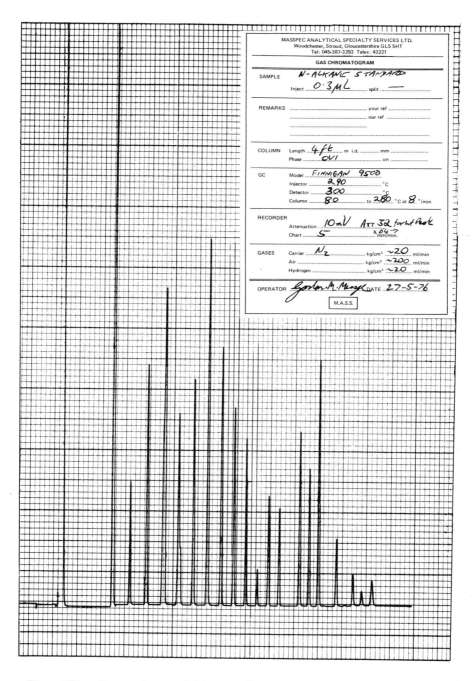

Figure 131. Operator's record ticket attached to output data encourages adequate record keeping. (Courtesy of Masspec Analytical Ltd.)

Sample Code *E 542/2.*   */100 ppm DINOSEB*   Date *19-10-81*

Sample Inlet Mode   *FORMULATION*   GC/~~ML~~/~~SP~~

G C CONDITIONS

Column *CPSil 5* *25 metres*	Carrier Gas *Helium*		Flow rate *1·0* mlmin^{-1}	
Inj. Temp *200* °C	Separater Temp *170* °C		Trans line *190* °C	
Column Temp *40* °C *for 1min*	Program *6 Degrees/min.*			
Capillary				
Grob Split Ratio mlmin^{-1}				
Total flow rate		Column Flow Rate		mlmin^{-1}

ANALYSER

Emission	*0·30*	mA	I E at 0 amu	*PROG.*	V
Collector	*−0·21*	mA	at 500 amu		V
I.V.	*—*	mA	Extractor	*3·2*	V
Electron Energy	*25*	eV	Lens	*−28*	V
First Mass		a.m.u.	Multiplier	*−1703*	V
Mass Marker		a.m.u.	Last Mass		a.m.u.

CI CONDITION

Reagent Gas	/		Mode	GC/MU/SP	
Source Temp	/	°C	Source Pressure		µ

OUTPUT MODE

Tape Number (R)	*019*	Tape Number (P)		
Disc Number	*6*	Sensitivity	*10^{-7}* AV^{-1}	Calibration *FC 43*

ACQUISITION PARAMETERS

Mode	RGC/~~SIM~~								
File	*FM 6003*								
Title	*E 542/2 100 ppm dinoseb formulation /50 ppm DBPE (IS) @ 6ml*								
Mass	*50-100 , 101-260*	(		) (		)			
Range	(		) (		) (			)	
Integration	*1*								
Time	MOR								
Seconds	*1*								
Per Scan									
Threshold	2 ~~or~~								
Run Time	*99*		(	) (	) (	) (	) (	)	

RE 80422 500 2/81 TBL

Figure 132.   Typical GC/MS instrument log referenced to the sample. This type of log not only shows the instrument operating conditions but also has references to the data system archive. (Courtesy of Freshwater Fisheries Laboratories, DAFFS Pitlochry, Scotland.)

systems a little bit of preparation can save a lot of writing, or typing in this case. It it is not possible to configure the system to present a table or columns to be filled in, it is still only necessary to enter the relevant data providing an agreed-upon order is followed. For data systems without log files, it is possible to use the file name and title or comments lines to encode information. The main file or data name that is used when data are stored should ideally give information relating to operator or sample originator and the date of acquisition.

For instance, G1420K could be **G**ordon's eleventh sample (**K** = 11) on day **142** of 1980,—Wednesday, 21 May, 1980. The sample name or title could be R03200-250,10,23 indicating **R**esidue work, **0V1** column 3 (further details in column log book), temperature programmed between **200** and **250°**C at **10°**C/ min, electron energy **23** V. Other data could be added providing there was space and the sequence was agreed. There is no need to record some parameters— mass range for example—as these will be given or at least implied by the data system printouts.

To be useful this type of record keeping must satisfy the needs of the instrument operators, and it is worthwhile spending a little time to consider the work of the laboratory and assign suitable codes. The decoder crib sheet should be readily available at the instrument and copies supplied to all operators and users of the GC/MS service.

Data Storage

There is no point in storing data if they cannot be retrieved at a later date. The foregoing comments about record keeping are therefore most important. Manual recordings should be filed for easy access. Ultraviolet recordings tend to fade if not kept in the dark or fixed by spraying with a protective film. The general shape of recorder outputs is also a problem; they tend to be relatively long and thin. One laboratory solved this problem by determining the normal output lengths of paper and ordering appropriately sized folders. The charts were kept flat and were easy to find and read months later.

On-line data system storage media is expensive because space is limited. Dumping from disk to tape is a reasonable way to store data. It is important that the data be saved in such a way that the record can be read back without destroying current data. That is to say, some tapes can only be written and read as whole-disk copies, and reading the tape destroys current data on one of the disks. On single-disk systems some consideration must be given to preserving current data before reading the tape. Even when files are written and read to and from tape individually, care should be taken with very long runs. It may not be possible to read these back when a disk is partially used, due to lack of space. Often file sizes can be reduced to ease the storage problem. Much of a GC/MS run consists of space between peaks. Using an enhance procedure greatly reduces the data storage requirements. Always examine enhanced data for validity before destroying the original, as the enhance algorithms can throw up spurious results for overloaded and poorly resolved or zeroed data.

Sample | GC — Vacuum Conditions | Interface

Run N.	Date	Sample	Vial Ref	File Ref	Column	Phase	Initial °C	Final °C	Rate °C	Pressure PSI	Solvent Ref sec	High Vac Torr	Injector °C	Sep Oven °C	Analyser °C	T/F line °C
032	17-6-76	n-alkane Std		STD1	Jaggi 20n OV1	OV1	80	265	4	14	55	8x10⁻⁶	300	240	65	250
033	17-6-76	Limestone extract	#1	LE11	-"-	-"-	-"-	-"-	-"-	15	53	-"-	-"-	-"-	-"-	-"-
034	17-6-76	Limestone extract	#2	LE21	-"-	-"-	-"-	-"-	-"-	-"-	55	-"-	-"-	-"-	70	-"-
035	21-6-76	n-alkane Std	0881	STD2	Jaggi 220 OV1	OV1	80	260	4	14	48	7x10⁻⁶	300	240	75	250
036	-"-	oil gills -bend	OSVI	CAS1	-"-	-"-	-"-	-"-	-"-	15	50	6x10⁻⁶	-"-	-"-	-"-	-"-
037	-"-	-"- - water	OSVI	CAS2	-"-	-"-	-"-	-"-	-"-	-"-	52	6x10⁻⁶	-"-	-"-	-"-	-"-
038	-"-	-"- - birds	OSA1	CAS3	-"-	-"-	-"-	-"-	-"-	-"-	50	4x10⁻⁶	-"-	-"-	-"-	240
039	-"-	Tank sample	TS	CAS4	-"-	-"-	-"-	-"-	-"-	-"-	53	2x10⁻⁶	-"-	-"-	-"-	230
		Aborted — leaking transfer line furnube														
040	22-6-76	Tank sample	TP	CAS5	Jaggi 20n OV1	OV1	80	260	4	14	52	6x10⁻⁶	300	250	70	240
041	-"-	-"- repeat	TS	CAS6	-"-	-"-	-"-	-"-	-"-	-"-	50	6x10⁻⁶	-"-	-"-	-"-	-"-
042	-"-	n-Alkane Std	SVY#1	STD3	-"-	-"-	-"-	265	-"-	-"-	49	-"-	-"-	240	75	-"-
043	-"-	Limestone extract	SVY#2	LE12	-"-	-"-	-"-	-"-	-"-	-"-	50	7x10⁻⁶	-"-	-"-	-"-	-"-
044	-"-	-"-	SVY#3	LE22	-"-	-"-	-"-	-"-	-"-	-"-	52	-"-	-"-	245	-"-	-"-
045	-"-	-"-		LE31	-"-	-"-	-"-	-"-	-"-	-"-	51	-"-	-"-	-"-	70	-"-
046	23-6-76	-"-	SVY#4	LE61	-"-	-"-	-"-	-"-	-"-	-"-	50	6x10⁻⁶	-"-	240	-"-	235
047	-"-	-"-	SVY#5	LES1	-"-	-"-	-"-	-"-	-"-	15	52	-"-	-"-	-"-	-"-	-"-
048	23-6-76	MeSt 100mg		Test1	-"-	-"-	-"-	240	6	15	51	6x10⁻⁴	-"-	240	75	240
	repeat	MeSt 200mg		Test	-"-	-"-	-"-	-"-	-"-	-"-	-"-	-"-	-"-	-"-	-"-	-"-
		Clean Source														
049	24-6-76	n-Alkane Std	AA1	STD4	Jaggi 20n OV1	OV1	80	265	4	15	48	5x10⁻⁶	300	245	70	240
050	-"-	Plant extract	AA1	AAPE1	-"-	-"-	-"-	-"-	-"-	-"-	50	-"-	-"-	240	-"-	-"-
051	-"-	-"- MugFrog	AA1	AAPE2	-"-	-"-	-"-	-"-	-"-	-"-	51	8x10⁻⁶	-"-	265	75	245
052	-"-	-"- -"-	AA1	AAPE3	-"-	-"-	-"-	-"-	-"-	-"-	48	-"-	-"-	-"-	-"-	-"-

(a)

MS Conditions

Run No	Pre-amp A/V	E.M. V	Lens V	Extractor V	Ion Energy V	Ion Vol mA	Collector mA	Emission mA	Comments
032	10^{-8}	1600	-45	1.3	69-85 70	0.09	0.41	0.30	
033	-	-	-	-					
034	-	-	-	-					
035	-	-	-60	1.5	76-10 69	0.15	0.15	0.30	
036	-	-							
037	-	-	-62	1.6	7.3-103 70	0.16	0.15	0.21	
038	-	-	-	-	-	-	-	-	
039	-	-						-	Pressure problems — leaking Transfer line jamube
040	10^{-8}		-72	1.7	8.0-12 70	0.15	0.15	0.35	Retuned
041			-75	1.5	8.1-12.1 67	0.15	0.14	0.29	
042	10^{-8}	1650	-70	1.4	8.3-12.4 70	0.16	0.14	0.30	
043	-	-	-	-	-	-	-	-	
044	-	-	-	-	-	-	-	-	
045	-	-	-	-	-	-	-	-	Sensitivity falls off
046	10^{-8}	1700	-72	1.5	8.4-12 70	0.16	0.14	0.30	
047	-	-		-	-	-	-	-	
048	10^{-8}	1700	-71	1.6	9(vppm) 70	0.16	0.14	0.30	low segment output width
									Clean In Source / Probe
049	10^{-8}	1600	-50	1.2	4-61 70	0.08	0.22	0.30	very lines
050	-	-		-	37-56 25	-	0.10	0.20	-
051	-	1700	-52	1.2	45-55 -	-	-	-	0.31
052	-	-	-	-	-	-	-	-	optimise by anal

(b)

Figure 133. Log book approach to GC/MS record keeping. (a) Sample and input conditions. (b) Mass spectrometer conditions. (Based on data supplied by Masspec Analytical Ltd.)

OPERATING PROCEDURES

The mechanics of GC/MS instrument operation can be learned only by using the system. Guidance should be sought from more experienced operators and from the manufacturer's manuals. (Note that this text is written as a supplement to the instrument manuals and is in no way intended as a replacement.) Operator manuals are written by people with a great deal of experience with the instrument, and even a user of long standing can often glean something new by rereading the manual. In many cases this means reading it for the first time, hence the old saying, "If all else fails, read the manual"!

Unfortunately, operator manuals tend to be written in such a way as to cover every point exactly in an unambiguous manner and therefore include a lot of detail. Even so, many facts may be assumed. Armed with some knowledge of your system, you can usually edit the operational procedures, adding notes and comments where appropriate to establish a comprehensive operating procedure. This is very useful for future reference, when you must train a new operator for example, and for infrequently used procedures.

Having done this, it is often worthwhile to go one stage further and prepare check cards for all routine procedures. These are especially useful in multi-user situations and as a guide for less-experienced operators and assistants. The idea of check cards can be extended to cover nonstandard situations such as unattended operation and emergencies. A notice displayed on the instrument explaining that the instrument should be running in a certain way will often prevent finger trouble from the casual passerby such as the boss or the cleaner. Brief and clear instructions of what to do in an emergency are also a good idea, and these should be placed on the instrument in full view when it is left unattended. Similarly, notices on main service controls, especially if they are not right next to the instrument, can prevent some disastrous shutdowns. A notice on the building's main water supply stop-tap such as

Do not switch off without reference to GC/MS laboratory personnel. Out of hours, contact _____

may prevent a leaking washer from becoming a leaking vacuum system when the local plumber comes to fix it. The same comments apply to gas and power supplies. The warning label on gas cylinders should specify the normal regulated tank pressure as well as the location of the gas users.

Operating procedures as described on check cards must be brief to be effective. Distinction must be made between those intended for the experienced user and those for the layperson. The plumber or the nightwatchman patrolling the laboratory may not understand even the most simple technical terms, as they will be outside his or her experience. Emergency procedures or warning notices must be unambiguous and in simple language to avoid confusion at a time of crisis. On the other hand, operator checklists can and should use the necessary technical jargon. Their purpose is to prompt and remind the operator, who should be able to refer to a full set of operating instructions if necessary. The full instructions may be part of the instrument manuals or written up as part of the protocol for each type of analysis. The check cards need not be confined to the GC/MS system and may relate to all aspects of the routine laboratory work.

SAMPLE PREPARATION

Samples will come from a number of sources, depending on the type of GC/MS laboratory. They may be generated internally or brought in for analysis by other sections, departments, or even outside customers. In all cases it is important to have a knowledge of the sample and its preparation. This is most easily achieved by asking the originator to fill in a sample analysis request form. These forms then constitute the basis of a sample log book. Operators in service laboratories find this approach very useful because customers often bring in a sample, saying "It is similar to the one you ran the week before last." The sample may be the next one in a sequence to the customer, but to the GC/MS operator it is effectively an unknown unless he or she can relate it to a previous sample. The sample request log will show not only the customer's expectations—because there should be a question relating to the type of analytical method, column, and so on required—but also the results achieved, with the operator's comments. The new sample can then be tackled with a greater probability of a successful outcome on the first analysis.

The condition of the sample is an important factor not only in terms of analysis but also in terms of GC/MS performance. Dirty samples, from biological or environmental extractions, for instance, can be difficult to analyze and soon contaminate the mass spectrometer. Whenever possible it is a good idea to review the sample extraction and workup procedure to take into account cleanliness as well as concentration. Sensitivity and detection limit are not the same thing. Detection may be seriously hindered by interferences, while instrument sensitivity may fall off rapidly due to contamination. Customer samples (that is, outside samples) are unknown in this respect. It is worthwhile to discuss instrument limitations and requirements with regular customers so that sample preparation can be optimized. High-quality reliable data are in everyone's best interest, and attention to these details during preparation can lead to better results. The GC/MS operator must recognize, however, that the customer cannot always provide a clean and pure sample. Some consideration should therefore be given to the amount injected.

It is important to establish the sample concentration and analysis conditions before using the GC/MS system. This can be done by running the sample in a stand-alone gas chromatograph first. Not only can the optimum column conditions be established, but with experience the response of the detector can be related to likely GC/MS performance. The relative concentration of the sample can be estimated and the sample can diluted as necessary. The chromatography traces could be provided by the customer, in fact, this is a condition of sample acceptance in some laboratories. Standards of chromatography vary widely, and ideal stand-alone gas chromatograph, parameters are not ideal GC/MS parameters. It is often desirable to run or rerun the sample on a stand-alone gas chromatograph with GC/MS parameters. Ideally the column used should be transferred from the gas chromatograph to the GC/MS. This approach to every sample can seriously slow down the analysis process. Many samples are repeats of similar work with known conditions and can be analyzed directly. It is very much up to the GC/MS operator to set the required standards.

Another important factor to consider when accepting samples for analysis is the solvent to be used. Many solvents are quite suitable for stand-alone gas chromatography work but not for GC/MS analysis. Chloroform, for instance, finds wide application in biological work but is not a good solvent for GC/MS as it tends to shorten the life of source and ion-gauge filaments. Other solvents such as benzene tend to hang around the vacuum system and take a long time to pump away. Try to avoid these undesirable solvents, and explain the importance of suitable solvents to the GC/MS customer.

AUXILIARY INLETS

Most GC/MS systems have additional inlets, either direct such as a solids probe or through an interface such as LC/MS. As with GC/MS, the mass spectrometer and its auxiliary inlet should be operated to achieve the best overall results. "Best" will depend on circumstances. With solids probe work, for instance, it is all too easy to scrape the sample into the probe cup and grossly overload the instrument. "If you can see it, you probably have too much" is a phrase worth remembering. There are times when a relatively large sample must be run, but not as a matter of routine.

Solids probes have vacuum interlocks associated with them that are used to remove the excess air before inserting the probe into the main vacuum chamber. The pumping lines to the interlocks are usually small and in vacuum terms relatively inefficient. Fortunately, the main vacuum chamber will usually accommodate a small blast of air. A compromise is necessary here. Extensive pumping of the probe interface will reduce the amount of gas entering the mass spectrometer, but the amount of backstreamed pump oil will increase. The contamination effect of this oil may mask the sample being analyzed if the oil is allowed to build up. One should pump down only as far as necessary, to the levels recommended by the system manufacturer but no more. To protect them, sensitive electronics and filaments should be momentarily turned off as the vacuum interlock is opened and the probe is pushed into place.

Some lubrication on the sliding seals is often beneficial, especially in the case of Viton 0-rings, but it should be used very sparingly to avoid contamination. Pumping out the interface can also cause sample loss from the probe due to evaporation. Probes often have cooling facilities to overcome this problem. To prevent excessive condensation on the probe, it should be secured in the interlock before cooling. The probe tip is then cooled just before pumping out the interface volume. There is only a small amount of air around the probe tip before cooling, so little condensation will occur.

Other inlets should be operated with the same attention to detail. For the occasional auxiliary inlet sample it is reasonable to leave the GC/MS interface operational. The source operating parameters may not be at optimum if there is a lot of helium carrier gas entering from the column, and if possible the flow rate should be reduced. If a lot of auxiliary inlet samples are to be run, it is advisable to disconnect the column and blank off the GC/MS interface so that the source parameters can be optimized for the inlet in use.

PLANNING

It is generally true that with practice one gets better at things. Hence, it makes sense to continue with similar work on the GC/MS system rather than chopping and changing. Set up times can be considerable and in themselves produce no data; however, they should not be rushed. Proper setup procedures are vital to successful and reliable results. However, if a batch of similar samples or samples requiring similar conditions can be run, the dead time due to setup procedures can be minimized. To organize the work load in this way requires some thought and planning as well as a prior knowledge of the samples.

Customers who want samples analyzed have time scales that vary considerably. People will often be prepared to wait a day or so for an analysis if it can be shown that the quality is generally better, that reliability is higher, or that the overall number of samples that can be run by the laboratory is greater when a particular mode or sequence of operation is adopted. It all depends on the individual situation. Obviously, customers should not be made to wait unnecessarily for instrument time. It is also important to keep them advised when delays occur.

Instruments also seem to perform better when operated in a consistent way. Part of the setup time for a new analytical method always includes the settling time of the system. There are often preferred sequences for analyzing a variety of samples under different conditions. Some laboratories find that EI work can be satisfactorily followed by CI work but not the other way around. In others, certain samples always foul the system and are best left until the source needs to be cleaned anyway or at least until the end of the day or week. This allows the instrument time to recover overnight or over the weekend. This sort of information cannot always be predicted but can often be obtained by careful study of the instrument log book after a few months of operation. Certainly, the operator should be aware of the possibility of interference between samples in terms of overall performance and should plan to minimize these effects.

The workload plan should also allow for instrument maintenance. Breakdowns cannot be predicted and should not be planned. In other words, time should not be allocated to create gaps in the weekly workload timetable so that there is contingency for failure. Failures never occur at convenient times; it is better to run as efficiently as possible with maximum instrument utilization and deal with trouble when it occurs. On the other hand, longer-term plans, especially budgeting, should allow a reasonable down-time percentage.

Routine and preventative maintenance, regular performance checks, and setup times can all be estimated. There is a great temptation to cut corners on these. Changing a disk drive filter, for instance, including cleaning around the data system, checking and cleaning fans, etc., could easily be half a day's work as part of the data system maintenance. In some laboratory environments this must be done every 3 months. The system may last for 6 months or even a year without failure, but disk drive head crashes, when they do occur, can be spectacular in more ways than one! It is not worth the expense and frustration just to get an extra half-day's work out of the system. Planning ahead allows one to be fairly rigid about these maintenance tasks. Exception can always be made for the very important sample that arrives at the last minute. Try to avoid letting these things slide, or they will never get done until they have to be, regardless of work requirements, because the instrument will be out of action.

OPERATORS

The more operators assigned to a GC/MS system, the greater the number of problems. One should recognize that the instrument is an extremely complex system and adequate operator skills cannot be achieved casually. No one should be allowed to use a GC/MS instrument without a basic level of training followed by a supervised probationary period in which it is established that the training has been successfully absorbed. After this, if there must be several operators, it is best if one person is designated the principal operator with the responsibility for the system and the authority to make decisions relating to its use. Even if all the operators are very experienced, this is an important point to follow. No matter how experienced they are, there can be only one captain of the ship. The captain is on the bridge, sometimes at the wheel, but never back on shore. The person in charge should be in close contact with those using the system and not someone in an office or laboratory elsewhere.

A procedure that works well in many laboratories with advanced GC/MS data systems is to allow the customers to process their own data at a second terminal. The basic results are analyzed and presented by the principal system operators. Subsequent or additional data processing is performed by the customer. This procedure can be extended as the customer becomes more experienced and is allowed to do some of the instrument operation, but the basic setting up and monitoring of the system should be done by the principal operators. This procedure is of benefit to the regular customer as well as to the busy operators of the GC/MS laboratory. As he or she gains some experience of the GC/MS instrument, appreciation of its capabilities and limitations also grows, and as a consequence the customers are likely to get more out of the system and achieve a higher quality of sample analysis.

CHAPTER TEN

Preventive Maintenance

A GC/MS system is a complex mixture of equipment encompassing a wide range of instrumentation techniques: electronics, vacuum, ion optics, chromatography, and so on. It would be unrealistic to expect the system to function without some sort of regular maintenance. Attention should be given to those factors that maximize instrument performance as well as those that prevent serious breakdowns. An instrument giving poor results may as well be out of service. The data will be unusable, at best suspect. Operating an instrument that is in poor condition is a waste of resources. Time, effort, and money spent on maintaining an instrument in tiptop condition pays dividends not only in the quality of results obtained but also in the morale of those associated with it.

SAFETY

Much of the routine maintenance work on a GC/MS system exposes the operator to potentially hazardous situations. There are high voltages, hot surfaces, and rotating machinery. Be aware of the dangers, and do not take chances. Only perform operations that you are qualified and confident to do. If possible, there should always be more than one person in a laboratory. Never undertake maintenance work on an instrument if you are working alone or allow anyone else to work unsupervised on an instrument.

Quite apart from the normal hazards associated with electronic equipment, GC/MS laboratories contain a wide variety of chemicals and samples that may be toxic, inflammable, or carcinogenic. Be mindful of the dangers to yourself and to others, especially visitors who may not be experienced in your working environment.

INSTRUMENT CARE

The GC/MS system should be kept generally clean and tidy. This means paying attention to areas behind and underneath the instrument as well as on top. Cooling air for fans, pump motors, disk drives, and so on, is often taken from behind or below the cabinets. Any buildup of dust and dirt is soon transferred into these units. Oil leaks from pumps are harder to pinpoint if the area under the instrument is already dirty. Rotary pumps are best sited in trays to minimize the effect of oil leaks, but the trays should not be allowed to fill up with oil. The trays can cause excessive noise because of vibration, which can be alleviated by standing the pump on a cork mat. Sound and dust levels can generally be reduced by standing the instrument on a carpet. One should be careful in the choice of carpet material to help avoid the effect of static voltages.

If possible, keep the laboratory environment clean, dust-free, and at an even temperature. Air conditioning is usually essential, as a GC/MS system produces a lot of heat even when the oven is off. A few degrees rise in the laboratory temperature can often push some of the internal GC/MS/DS units to near their maximum ratings. Component life, and therefore the time between failures, will be enhanced by a cooler environment. A cooler laboratory encourages the wearing of laboratory coats and is also good for operator morale. Working efficiency is better providing the siting of the air-conditioning unit does not cause excessively cold drafts over the work area. Many otherwise excellent air-conditioning or laboratory cooling units are switched off for this reason.

Each instrument has a characteristic sound and appearance when all is well. For instance, one can often sense something is wrong with the vacuum system by an audible change in the pumping system and a change in the brightness and color of the ion-gauge filament. Cultivate an awareness of your instrument. Replace blown bulbs in switch indicators and display panels as soon as possible so as to maintain the ability to confirm instrument status at a glance. Apart from indicating bulbs, other items wear out or become contaminated with use. All cooling fans should be checked once a week to confirm that they are rotating. Clean their grills and filters if necessary.

Rotary and selector switches become electrically noisy with use; an occasional drop or spray with a proprietary switch cleaner and lubricant will improve their operation. Some potentiometers will be subject to more use than others and should be replaced if they show excessive electrical noise or dead spots due to wear. Try to avoid touching display screens, which become scratched and etched with body oils and acids. When the display definition is reduced, operator fatigue and eye strain is increased. Displays usually have a front screen, which is a relatively inexpensive component to replace if they become badly affected. In some cases it is possible to polish out light marks. Refer to the operator's manuals for details.

Avoid placing samples and solvents on the instrument. Quite apart from spoiling the paintwork if they are knocked over, many solvents will ruin plastic materials such as switches, knobs, and display screens. There is also a serious fire risk with some solvents when they are close to hot surfaces. Try to arrange a work surface near the instrument to be used for syringe rinsing and sample preparation.

From time to time the operator must disassemble or adjust parts of the instrument. Try to get the right tools for the job so that screw and bolt heads are not burred. When they are hot, some fittings will expand so much that the "correct" size spanner will not fit. If possible, have a set of spanners opened up by grinding one face just enough to cope with this situation, bearing in mind that they may then be unsuitable for cold operation. Usually only a few sizes are involved, and so having the two versions available (clearly marked) does not cost a great deal.

In addition to the basic hand tools, a reasonable quality multimeter is a useful addition to maintenance equipment so that power supplies and other test points can be monitored. A test meter is also essential on some instruments for routine setup procedures. On balance, an analog meter is probably preferred to a digital meter. Analog meters usually have a higher voltage range and are more suitable for tuning measurements where small trends or dips in response must be observed during adjustment. A more expensive digital voltmeter can be purchased if more extensive maintenance work is to be carried out.

System Checks

It is nice to know that everything is running as it should. The daily check of the instrument pressures, sound, background, peak shape, and resolution will generally be sufficient for the experienced operator to know that all is well, but some prefer the added confidence of knowing that the electronics is exactly adjusted. The operator's service handbooks usually give the location and specification of all the electronics test points. The problem with checking only test points is that instruments vary a great deal and the specification limits may have been laid down in the manual by an electronics designer with little regard to overall power supply loading and the point of measurement, not to mention production tolerances. Logic power supplies on some instruments routinely run at 4.8 V, although the specification is 5 ± 0.1 V. Is this a fault or not? It is very difficult for the operator inexperienced in electronics to decide.

Certainly you should check test points from time to time, especially if they are readily accessible or have been uncovered due to some other work on the instrument. Seek advice from a service engineer if the voltages are outside limits, but do not be ruled by the stated limits. If things are going wrong, they should show up as faults on the normal displays or in the data. Test points can then be used to pinpoint the trouble.

When working on the design of very stable electromagnet power supplies for the Micromass 70-70, we were looking for stabilities of better than 10 ppm in a current of about 6 A. Electronically this is a very difficult thing to measure without introducing noise and instability into the system. Tuning onto the side of a peak told us if the magnet (and a lot of other things) were stable. The mass spectrometer itself was our measuring instrument.

Performance checks in terms of sensitivity and detection limits are a different matter. One does not always run the instrument at its limits, but for a certain sample it may be necessary. The overall performance varies with the condition of the source as well as with the lifetime of some of the GC/MS components. Hence it is of value to run a set of standard examples under standard conditions as part of the routine work of the laboratory. The frequency with which this should be done will depend on circumstances, but at least once a month is recommended. The standard samples used should be relevant to the work of your laboratory and need not be those used by the manufacturer to define specifications. Performance with methyl stearate is a good general indicator, but a very low-level DDE solution may be more appropriate for a laboratory doing pesticide residue work.

Spare Parts

A GC/MS system consists of a large number of different parts. Some have a relatively short life in use. In addition, a variety of items are used in the process of recording data. These consumable items are trivial if you have them to hand but assume a significance out of all proportion if they are unavailable when needed. A small stock of items such as ferrules, septa, and O-rings should be stocked and replaced as used. Similarly, quantities of recording paper, recorder ink, and pens should be maintained.

Often significant discounts are available to bulk purchasers of recorder paper. With plain paper this is all right, but photographic papers such as those used in UV recorders and electrostatic printers as well as those used in thermal printers have a limited shelf life. The paper's shelf life can usually be improved by keeping it cold. These papers have been stored in the refrigerator for quite long periods satisfactorily. If the paper is refrigerated, it should be allowed to come to room temperature before use, or condensation may occur inside the recorder. It is advisable to have at least one spare roll or pack in the laboratory ready for use. Of course, the paper should also be protected from moisture when placed inside the freezer or cold storeroom.

Filaments are liable to blow at any time. In some cases they can be repaired, either by the user or by a commercial service, but it is more convenient if there is a spare available for immediate use. Many users prefer to allow for one in the instrument, one ready for use, and one for repair. Like the mass spectrometer filaments, ion-gauge filaments are also liable to fail. In many cases they cannot be easily repaired and a replacement gauge must be used. The mass spectrometer electronics are usually protected by circuits that monitor the system pressure, using the ion gauge in the metering circuit. The system should not, and in most cases cannot, be operated without this protection, and so a spare gauge is vital.

The items just described are often referred to as "consumables" by the GC/MS manufacturer, and certainly they are consumed as the system is used. Other items are similarly described in spare parts price lists but are more correctly described as maintenance parts. These are parts that are used only during routine work or for emergency service. If the maintenance of the instrument is carried out by its users, then some of these items will be needed on a routine basis and it is advisable to carry them as spares to avoid purchasing and shipping delays. It is good practice to have a set of spare O-rings, pump oils, bulbs, filters, and so on, anyway. Downtime on the instrument can then often be used to good advantage to do a routine maintenance job. One should not delay routine maintenance until a fault occurs, but taking advantage of a slack period or an unscheduled interruption can reduce downtime later by bringing forward any planned maintenance.

Unexpected downtime sometimes occurs due to the failure or accidental damage of a critical item. The holding of items such as separators, gauges, filaments, and multipliers can be considered an insurance policy against loss of instrument time. Shelf life for these items may be important. For instances, continuous dynode–type multipliers have a long shelf life because their tin oxide surface coating is not susceptible to degradation in air. On the other hand, box-and-grid or venetian blind types with their copper—beryllium surfaces are oxidized by air and water vapor and become unusable. Storage conditions are very important if the critical item is to be useful after one or two years on the shelf.

The cost of spares for GC/MS systems is a common bone of contention between users and the GC/MS manufacturers. Often groups of users can effect savings by buying quantities of components directly from the component manufacturers. GC/MS system manufacturers do not usually make a lot of money out of spare parts and are often willing to give names and addresses of alternative suppliers, since their business aim is to sell GC/MS systems and not spare parts. Nevertheless, spares can be expensive. A wise department will make adequate budgeting provision for running and maintaining its instruments.

PREPARATION AND CARE OF COLUMNS

Fundamental to good GC/MS work is good chromatography. This can be achieved only with a column that is suitable for the analysis and that has been well prepared and maintained. Empty column tubes should be cleaned and conditioned before packing. This can be done by heating them in a gas chromatograph oven. One end should be connected to an injector assembly. The other end should be supported to prevent damage to either the column or the oven. The column should then be heated to at least 20 or 30 degrees above its expected maximum working temperature with a flow of helium or nitrogen through it. The minimum temperature to be used depends on the subsequent packing material. Some people prefer to condition all empty column tubes at the same temperature regardless of the packing to be used and set aside an old gas chromatograph for this purpose, often doing several tubes at once. This cleaning and purging process should continue for at least 4 h. Overnight is better. The tubes should be cooled while the gas flow is maintained and then filled with packing material.

A practical advantage of the purging process is that the inner walls become very dry and it is easy to get the packing material to move down the tube. This is an important consideration with coiled columns. The packing should fill the column without any gaps or loose areas, but one must take care not to overpack and crush the packing or flow problems may occur. A small funnel with a short joining piece of tube is essential. The packing can then be tapped or shaken into the column or, better still, sucked in with the aid of a modest mechanical vacuum pump or a faucet-mounted aspirator pump. With the latter, one should take care not to allow the pump to blow back, which would soak the column and its contents.

A plug of sylanized glass wool is normally used at each end to prevent the packing from falling out. The sylanized glass wool and packing materials can be purchased from most chromatography suppliers. Ready-packed columns can also be purchased, but the incidence of glassware damage from shipping is high and so empty tubes can often be made locally more cheaply. Some people also prefer to prepare their own packing materials. This is quite acceptable, but the inexperienced in this art—for an art is certainly is—are advised to experiment with stand-alone gas chromatographs first, where the problems of instrument contamination are less severe.

Once packed, the column should be conditioned. Initially it should be connected in an oven with a carrier gas flow of say 20 mL/min and warmed slightly to, say, 60 or 70°C. These conditions should be held for about an hour, during which time most of the air will be blown out of the column. Try to avoid excessively heating the column when it is full of air, because there is a danger of oxidizing the packing material and any impurities that it contains. The oxidation products are likely to cause active sites and degraded column performance.

Once clear of air, the column should be programmed slowly at only a few degrees per minute up to the maximum operating temperature for the phase. The maximum temperature should be held for a couple of hours to allow the phase to spread. One should not condition for very long periods at the maximum operating temperature, or the phase may be stripped off the support material, leaving absorbing active sites. Obviously, at elevated temperatures the bleed level will be high, and one should not program the column with its exit connected to the mass spectrometer interface. After conditioning, the column should be cooled slowly while maintaining the gas flow.

Although the column is now ready for general use, it is a good idea to run a standard test substance as the first sample. A compound or mixture appropriate to both the column phase and its intended application should be used. The response should be checked carefully for chromatographic performance in terms of resolution, retention times, and number of theoretical plates. Watch for tailing and spurious peaks, which indicate active sites, contamination, and/or dead spaces due to variations in packing density. Watch also for excessive bleed, which indicates insufficient conditioning time. The data should be saved and the column suitably labeled so that its later performance can be compared with that at the time of making. The data should also be compared with data from similar columns to check for consistency of the preparation technique.

If the column is not to be used immediately it should be disconnected from the oven and stored. The ends should be plugged to prevent the ingress of dirt and absorption of air and water and other airborne contaminants. This also applies to any column, old or new, that is not in current use. The columns should be clearly labeled to identify their size and packing and their particular data file or entry in the column log book. The above discussion relates to newly made columns, but from time to time it is also worthwhile to thoroughly condition an old column so as to restore some of its lost performance.

If the making of packed columns is an art, then some say that the making of capillary columns is a black art. It may not be magic, but certainly a lot of know-how and experience are needed to make even reasonably good capillary columns. Most laboratories prefer to purchase ready-made columns from the experts; even so, one should not assume that the column is perfect and ready for use. The columns usually come with a test chromatogram that may not be of the compounds of interest. It is advisable to run a standard sample of a relevant compound or mixture to check the column in your system. Figure 134 shows a GC/MS trace of a typical hydrocarbon test mixture. It consists mainly of the normal alkanes, containing between 14 and 36 carbon atoms. There are a few impurities, especially in the early scans. The column resolution is demonstrated between scans 230 and 290. Here, normal C-17 alkane is satisfactorily resolved from pristane, and normal C-18 alkane is resolved from phytane. The trace has been thresholded, that is, brought to zero at scan 550, so that the relative bleed levels across the range can be clearly seen. These are quite low and are acceptable, providing that the absolute level, which is not shown, is also satisfactory. The column was tested with this mixture prior to use in hydrocarbon survey analysis.

Capillary work often involves a wide range of unknown compounds. A more general test mixture is therefore preferred if the column is to be fully evaluated and tested. Figures 135 and 136 show typical GC/MS response to test mixtures similar to that in Table 6. The test mixture in each case was based on the recommendations given in a number of papers and lectures by the Grob family. The response to all the compounds except the acid is satisfactory. The acid is a very difficult compound with which to get satisfactory chromatography. The response to the other components is generally good, and the columns are therefore useful for a wide range of compounds.

Loss of capillary column performance is often due to stripping of the phase at the injector end. Turning the column around from time to time helps. If necessary, a short length can be cut off the end to remove the defective section. This often happens as a matter of routine with glass columns due to accidental breakage! Fused silica columns with their polyimide coating do not break so often, although they do tend to become brittle when exposed to excessive heat, such as in the injector assembly and through the transfer assembly. It is advisable to condition the very flexible columns for a time before connecting them to the mass spectrometer, where they are directly coupled to the ion source. There is a tendency for the column's outer coating to bake off and varnish the source.

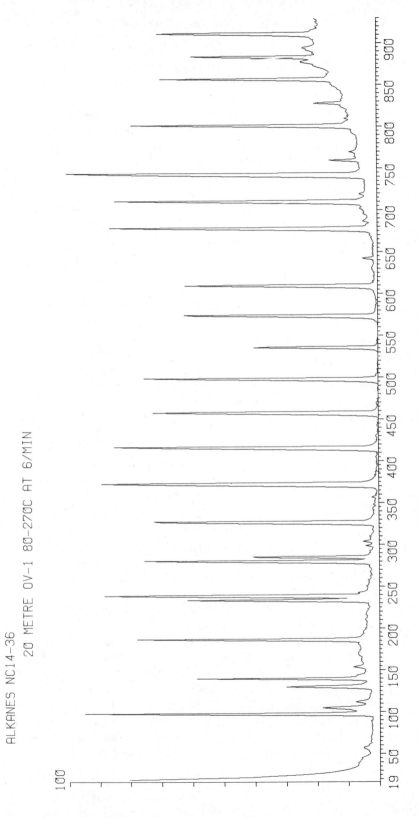

Figure 134. GC/MS RIC trace of a capillary column hydrocarbon test mixture.

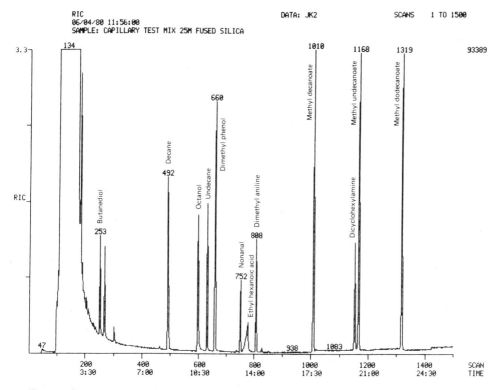

Figure 135. RIC trace of a general-purpose capillary test mixture. (See Table 6.) (By kind permission of Finnigan—MAT Ltd.)

TABLE 6
Typical Capillary System Test Mixture

Compound	Concentration, mg/L
C_{12}—acid methylester	40.8
C_{11}—acid methylester	41.2
C_{10}—acid methylester	41.7
Decane	28.3
Undecane	28.7
1-Octanol	35.5
Nonanal	40
2,3-Butanediol	40
2,6-Dimethylanaline	32
2,6-Dimethylphenol	32
Dicyclohexylamine	31.3
2-Ethylhexanoic acid	38

259

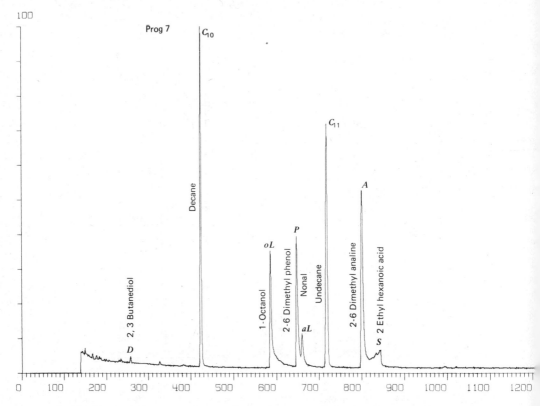

Figure 136. RIC trace of a text mixture similar to that shown in Figure 135 and Table 6. The peaks shown correspond to the following components (with approximate scan number): 2,3-butanediol (260), decane (420), 1-octanol (590), 2,6-dimethyl phenol (650),

Heating the column for an hour or two in the oven before connecting it to the mass spectrometer reduces this effect.

Ideally a conditioned column will give satisfactory performance right from the beginning, but in practice it may be necessary to condition it further by injecting a relatively concentrated sample. Some of the sample will be absorbed onto the active sites, and the performance will improve with each injection. Blanks should be run subsequently when doing very low-level work to make sure that the absorbed sample is not released. Columns tend to improve when they are used for the same type of sample, providing the sample is clean. It is advisable to keep several columns of the same phase, one for each type of analysis, so that the optimum performance can be achieved for each type of analysis.

Closely associated with the columns are the injector septa. Some people also like to condition the septa by baking them in an oven. This technique is successful in reducing the septum bleed, but the septa do become harder and less plia-

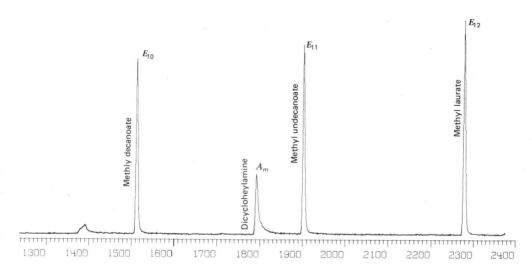

nonanal (660), undecane (720), 2,6-dimethyl aniline (800) 2-ethylhexanoic acid (840), methyl decanoate (1510), dicyclohexylamine (1790), methyl undecanoate (1905), methyl laurate (2280). (Courtesy of Freshwater Fisheries Laboratories, DAFFS, Pitlochry, Scotland)

ble and so their working life is reduced and they are more likely to cause syringe damage. If a septum is causing a particular interference it is sometimes better to change the brand or type rather than to bake it hard and risk needle damage or sample loss due to septum leaks.

Sylanizing Reagents

Try to avoid the use of sylanizing reagents wherever possible. Careful preparation of columns with high quality materials should give satisfactory results. Sylanizing reagents will coat the mass spectrometer components and greatly reduce their performance. In the case of multipliers this effect may be permanent. Even injecting the reagents into columns that are not connected to the interface can cause problems later, as some of the material will be retained temporarily in the column, later bleeding out into the mass spectrometer. If column performance is degraded so much by the dirty samples that sylanization appears necessary, then give some thought to improved methods of sample cleanup. One solution is to have a special column tube made with a wider bore for the first 100 mm or so. Into this is fitted a removable insert that is packed in the usual way. Most of the contamination is caught by this small precolumn, which can be replaced at regular intervals so that consistent high quality results can be obtained. This technique cannot be used with capillary columns.

INTERFACES

In an ideal world, interfaces would be all-glass, since glass has good chromatographic properties. All-glass systems can be made, but they are very fragile. Glass-lined tubing (GLT) is usually used to connect the column to the separator or flow restrictor and to interface with the mass spectrometer vacuum system. Great care should be taken not to bend GLT accidentally or twist it as the glass lining may crack, exposing the metal and adhesive support, with disastrous chromatographic consequences.

The interface assembly is normally kept hot but usually assembled cold. Allowance should be made for expansion of the fittings and mounting brackets. The assembly bolts should not be too tight to allow for some movement during warmup. It is advisable to put a small amount of high-temperature lubricant on the mounting bracket bolts to prevent seizure, but do not use the lubricant on the gas-line fittings, for it might get into the chromatographic system. Avoid metal ferrules on the gas lines. These swage onto the pipes and cannot be removed without damage to the assembly. For high-temperature work on fittings that are not to be made and remade often, that is, all of the interface assembly, Vespel ferrules are perferred. Graphite-loaded Teflon is also a useful material, but do not use pure graphite anywhere near jet separators or fine-bore transfer lines as graphite particles can so easily block them.

Monitor system pressures and diverter operation to establish normal operating characteristics. Changes in these are indicative of a partially blocked separator or transfer line. Do not delay in attending to a problem or the partial blockage may become permanent. Take care to avoid chipping glass tubes such as the separator stems when tightening up fittings. The glass particles can be another cause of trouble. Most interfaces are fitted with filter frits, which should be replaced regularly or the dirt that builds up will become an active surface. Take care not to crush the frits during assembly if they are at all fragile.

It is common, when disconnecting a column, to switch on the diverter in the belief that it will reduce air flow into the interface. In practice the transfer assembly from the oven to the interface is an insufficient restriction, and air flow through it is increased by this technique. As a consequence there is a greater likelihood of drawing dirt and dust into the separator.

As with columns, from time to time it is necessary to condition or, to be more precise, deactivate interfaces. This is especially true with platinum capillary interfaces. These should be disconnected from the instrument and thoroughly rinsed with methylene chloride. The platinum tube should then be rinsed with a methylene chloride solution containing 0.1 to 0.5% of either benzyldiphenyl phosphonium bromide (or chloride) or di-isobutylphenoxy ethoxy dimethyl benzylammonium chloride. The first compound is especially recommended for work with nonpolar compounds; the second is suitable for both polar and nonpolar compounds. Interfaces in general and platinum in particular can be further deactivated by coating with Carbowax 20M. Connect a column of Carbowax 20M to the interface and heat the oven and interface to about 130°. Leave overnight with helium flowing. This should deposit a thin layer of carbowax, coating the entire interface. The technique is not suitable for jet separators, which can easily be blocked by the liquid phase. These should be cleared and deactivated with commercial cleaning compounds such as a Decon solution.

The Carbowax treatment is carried out at a temperature above the normal operating point of the column and causes a high-level bleed which coats the interface. The coating will contaminate the ion source if it is not held at a relatively high temperature. Generally it is best to perform this operation just before a routine source cleaning. It should also be noted that this procedure is not good for the Carbowax column and an old but clean column should be used.

ANALYZER CLEANING TECHNIQUES

In use, the various parts of the GC/MS system become contaminated. To a certain extent the pumping system is self-cleaning and the buildup of contaminants in the analyzer is a relatively slow process. Unfortunately, a very small amount of contamination in the wrong place can drastically reduce instrument performance due to charging effects. In particular, the use of chemical ionization, biological extracts, and derivatizing reagents as well as the normal levels of column bleed all add to this problem.

Frequent opening of the vacuum system, for solids probe work for instance, tends to give a high level of pump oil background, which can start to mask the sample spectra. This usually clears if the pumping system is left to recover, say overnight, and with care is not normally a problem. If the interfering background persists or is unusually high (someone forgot to close the valve?) or for other reasons, it may be necessary to bake out the vacuum system. This involves raising the manifold temperature to about 250 to 300°C for a few hours to drive off the source of background contamination from the surfaces.

The instrument cannot be used during this process, so it is often done overnight under automatic or semiautomatic control, the heaters being switched off after a set time period. In principle, this sounds like a fine idea, but in practice it is often quite the reverse unless high temperatures can be maintained for fairly long periods. The problem is that the analyzer within the vacuum chamber is usually cooler than the outside walls because of poor heat transfer. As a result, condensation occurs and the contaminant moves from the walls to the analyzer.

This effect is generally more pronounced in the relatively smaller vacuum systems of quadrupole instruments and on any system if insufficient heating tapes and jackets are used.

It is good practice to keep the vacuum chamber warm at all times to reduce contamination. The actual temperature limits will depend on the operational characteristics of the instrument. The Finnigan 4000, for instance, has a multiplier with an upper temperature limit of 100°C when powered. In normal operation this means that the system can be operated with an indicated temperature on the manifold of up to 150°C due to the heat losses at the source and analyzer flanges. Most users run the system with a temperature set in the range 120 to 140°C.

If a bake-out is required to reduce the background level, it is a good idea to remove as much of the ionizer and analyzer as possible. On sector instruments it is not practicable to remove the analyzer section, as the flight tube is also the vacuum envelope. Electrostatic analyzer plates and the various slits can be removed, but it is a major undertaking. The flight tube is usually sufficiently open that a good heater arrangement can prevent significant condensation. The ionizer, which is a more enclosed unit, can be removed. The more compact ion source and analyzer arrangements on quadrupole instruments are easily removed. The vacuum ports should then be sealed with blank flanges and the system pumped and baked. The removed items should be cleaned separately.

After baking, the system should be vented with a dry clean gas, preferably helium. The cleaned ion source and analyzer sections should be reinstalled and the system pumped down. There should be no delay in getting the system back under vacuum. This procedure is normally effective in dealing with background contamination problems, and only in extreme cases is it necessary to strip and clean the entire vacuum and pumping system.

Source Cleaning

One of the regular chores in the GC/MS laboratory is source cleaning. The frequency of this task depends on the type and number of samples analyzed as well as on the analytical method, CI sources tending to contaminate more quickly. To minimize operational downtime, some laboratories have a spare source, or sources if different types are used. One source can be cleaned at a convenient time while the other is in the instrument. It is worth noting that cleaning one source is a chore, cleaning two is beyond a joke. The spare source should therefore be cleaned soon after taking it out of the instrument, and it should be cleaned thoroughly. Once clean, it should be wrapped in tissue and stored in a dry place ready for use.

It is also good practice to clean a source, even if it does not appear to be dirty, when it is only out of the instrument temporarily. This situation occurs with instruments where different sources must be used for different ionization modes. The partially dirty source will be coated with a very fine layer of material, especially carbon, from the fragments of ionized compounds. This layer will be extremely active and may be capable of absorbing considerable amounts of water and other compounds from the atmosphere. The source may not have the same characteristics when reinstalled in the mass spectrometer, because the nature of the surface coatings may change considerably. Obviously, there are times when one does not want to clean the source. It may be out of the instrument for only a few hours, or it may have recently been cleaned. To minimize the effects of absorption the source should be placed in a clean, dry container such as a desiccator or a jar purged with dry nitrogen or helium when it is temporarily removed from the instrument.

There is an art to cleaning sources. It is not always a matter of speed; as with most things, this comes with practice. It is more a matter of care and attention to detail. The basic principle involves removing the surface coatings with a mild abrasive, which can be in the form of a rouge paper, very fine grit or powder, or even a proprietary metal-cleaning paste or solution.

Abrasive papers are easy to use on large areas and flat pieces. They are ideal for flat plates and slits, especially in the larger high-voltage ion sources. They are difficult to use on smaller pieces, on the inner edges of fine holes and slits, for example, and inside chambers, such as ion volumes. In these circumstances, fine-grade grinding powders such as aluminum oxide are preferred. The powder is made up into a slurry with clean water. The addition of a very small amount of industrial detergent, one that is uncolored and unperfumed, helps to remove the dirt and improves the slurry by reducing surface tension. The detergent should be diluted considerably so it will not become a source of contamination. The pieces can be cleaned by rubbing with the slurry using a cotton-tipped stick. Wooden sticks are preferred; they can be split to give fine points and rough edges to get into awkward corners and holes. The wooden end can be used to attack any stubborn deposits, and the stick can be drawn through holes and slits to clean their inside edges. The nearer a surface is to the ion beam, the more critical it is.

It is important not to let the abrasive dry, as it is then often difficult to wash off. A good idea is to have a small beaker of water to dip the pieces in during cleaning to prevent them from drying. Again, the addition of a tiny amount of detergent helps. Once cleaned the pieces can be dropped into the beaker prior to rinsing. During disassembly of the source and the initial cleaning it is not necessary to wear gloves providing the abrasive cleaning stage is followed by a satisfactory degreasing process. On the other hand, one should never directly handle any pieces such as filament assemblies and ceramic parts that are not going to be cleaned later.

Once all the pieces have been scrubbed and placed in the beaker of water, they should be shaken for a minute or two in an ultrasonic bath. The beaker containing the parts can be placed in most small ultrasonic cleaners, but care should be taken to see that it does not float and topple over in the larger, deeper cleaners. The parts should then be transferred with tweezers into a beaker that contains distilled water with no detergent. Check that the parts do not have cleaning powder sticking to them. If they do, rub the powder off with a clean, wet cotton bud and give the piece an extra rinse. At this stage, it is a good idea to include the tweezers, pliers, and small screwdrivers that are used for source assembly in the cleaning process. These tools should not be used for any other job and should be kept as clean as the source. Avoid plastic-handled tools for this work; if they must be used, do not include them in the solvent rinses.

Rinse the parts ultrasonically in distilled water, acetone, and finally methanol. Always lift the pieces from one beaker to the next, using four clean beakers in all. Some laboratories find it useful to keep a set of beakers of the right size just for this purpose. Always check carefully for screws and washers that may have shaken loose. Finally, place the pieces on some clean tissue to dry. Methanol evaporates more slowly than acetone but leaves less residue. Take care that the tissue is on a clean surface, or contaminants may be leached onto the freshly cleaned pieces. In fact, a good technique is to wipe the pieces on a wad of tissue to remove excess solvent and then place them on several layers of crumpled tissue so that they do not rest on the bench surface.

Where possible, the ceramic spacers and insulators should also be cleaned. The smaller washer-shaped spacers can be threaded onto a stainless steel wire and then cleared of organic material by heating to a bright red, almost white, heat in a small flame. Always wear safety glasses when doing this in case a piece shatters. Hang up the wire with the spacers to let them cool. Do not lay them on the bench.

To remove metal deposits and to clean larger ceramics, which cannot be heated for fear of cracking them, use an acid wash. Soak the parts in a beaker of aqua regia for a few minutes to dissolve away any deposits. Then rinse them thoroughly with water. Some ceramic materials are quite absorbent, so thorough rinsing means flushing the parts with water for several minutes. This can usually be achieved by placing the beaker under a running tap. The ceramic should then be dried with a clean tissue and placed in a warm area to dry completely, a glassware drying cabinet being a suitable place.

Once the ionizer parts are clean and dry, they should be reassembled. Great care should be taken to keep the pieces clean; wear gloves. The type of glove depends on individual preference. Plastic (PVC) and rubber gloves do not breathe, and your hands become quite sweaty after even a short time; however, the contamination of the pieces is small. Some plastic or rubber gloves can give off quite a lot of phthalates (plasticizers), but these usually boil off the source fairly quickly under vacuum. Alternatively, woven cotton or nylon gloves can be used. They are more comfortable to wear but do not give a perfect barrier to the skin. A clean pair of gloves should be used each time the ion source is assembled, the gloves should be washed at regular intervals.

Proprietary metal polishes are an alternative to abrasive slurries made with grinding powders and water. Those designed to clean chrome or alloy and sold by most auto supply shops seem to be particularly effective in removing source contaminants. Unfortunately, these fluids, especially the creams, contain a significant proportion of greases that are very difficult to remove and can themselves be a cause of contamination. Very thorough washing with several solvent rinses is usually necessary to remove the cleaner. In some cases trichloroethane is the most effective solvent; in other cases, it is methylene chloride or one of the Freon degreasers. Certainly, one should experiment to find the most suitable solvent, if these polishes are to be used successfully. After the primary degreasing, the parts should be cleaned ultrasonically in acetone and then methanol as described earlier. Probably far more effective than metal polishes is the application of electropolishing to ion source cleaning.

Electropolishing

This is a relatively easy way of cleaning small ionizer components, although it is limited in effectiveness with complex shapes and inside chambers unless special probes are made. Electropolishing does not produce a scratched surface like the abrasive methods described above, and in fact it gives an improved surface quality that is often easier to clean subsequently. In many cases the smooth surfaces show a reduction in the buildup of dirt deposits.

Electropolishing sounds like the answer to the GC/MS operator's prayer, but there are limitations to its use and effectiveness. Not all materials will electropolish satisfactorily, and one should consult the instrument manufacturer before using the technique. The process removes metal from the surface, and so parts should be treated for a minimum length of time.

Surfaces that are shielded from the current flow, either due to their geometry or because of insulating deposits, will not electropolish, and an abrasive technique must be used first. Some operators find that a combination technique is useful. Dirty parts are connected to the cleaning apparatus and dipped into the electrolytic solution for a few seconds. Some of the deposits will be removed by the process. The parts should then be rinsed under a running tap and the remaining deposits attacked with an abrasive cleaning method. The initial electrochemical process seems to help reduce the adhesion of the deposits, and subsequent cleaning is often easier. When the major deposits have been removed the parts should be rinsed and given further treatment in the electropolishing process. In this way, cleaning can be both quick and effective, although one should not be in too much of a hurry. If the current is too high, there may be significant outgassing and local heating, giving rise to localized pitting of the component surfaces.

A description of the materials and apparatus used for electropolishing stainless steel, based on a method developed by Mr. Ben Fry of Finnigan Ltd. and used successfully by many Finnigan GC/MS users follows. Although the procedure describes the cleaning of the Finnigan model 4000 ionizer, the techniques apply equally well to most manufacturers sources.

Electropolishing Cleaning Method for Ion Source Cleaning

CAUTIONS: **The solutions recommended are extremely corrosive, and it is the responsibility of operators to protect themselves and colleagues by following accepted safety procedures. Before risking very expensive components, such as** complex ion volumes, experiment with some pieces of scrap stainless steel or one of the less expensive ionizer parts. Electropolishing solutions vary in their effectiveness depending on their age and the amount of use.

Materials and apparatus

1. Electropolishing solution suitable for stainless steel
   5 parts concentrated orthophosphoric acid
   3 parts concentrated sulfuric acid
   2 parts glycerol
   This solution is best made up by adding the glycerol to the phosphoric acid, using a magnetic stirrer to mix them well since both components are viscous. Finally add the sulfuric acid. Formation of this solution is exothermic, and therefore suitable Pyrex glass or plastic vessels should be used. This solution is initially a pale yellow that with repeated use will become dark green. Store the solution in a safe place, properly labeled.

2. Dilute nitric acid, approximately 10% aqueous

3. Methanol for final rinsing

4. Power supply. An inexpensive 6–12-V battery charger is quite satisfactory. Current flow is usually limited by the dimensions of the electrodes.

5. Electrodes. In the electropolishing process the workpiece forms the anode and is usually connected to the power supply using a stainless steel crocodile clip. The cathode should be of stainless steel wire or tube bent to form a circle in the bottom of the electrolyte vessel (Fig. 137). The vertical connection should be sheathed in insulation resistant to the acid.

The ion volume of the Finnigan model 4000 requires a special cathode that has an insulated end to locate in the hole used for the solids probe access, with about 10 mm of exposed stainless steel inside the ion volume. The remainder of this cathode is sheathed in insulation except at the end where the electrical connection is made. The details are shown in Figure 138.

For cleaning small pieces such as lenses, a magnetic stirrer will provide suitable agitation of the solution. For cleaning inside surfaces, a 20-ml glass syringe with rigid plastic tubing attached has proved ideal for circulating electrolyte inside the ion volumes and other small apertures.

Procedure.   For cleaning small components, the ring cathode is held near the bottom of a 250-ml beaker above a magnetic stirrer bar (PTFE or polypropylene-coated for acid resistance). The vertical connection to the cathode is insulated to prevent the development of an uneven field. Each workpiece (lens, etc.,) becomes the anode and should be connected to the power supply via a stainless steel crocodile clip. With this apparatus a current of 1 to 2 A is achievable at 6 V (12 V is not recommended, as effervescence becomes excessive). Stirring is necessary to disperse gas bubbles formed at both electrodes as rapidly as possible;

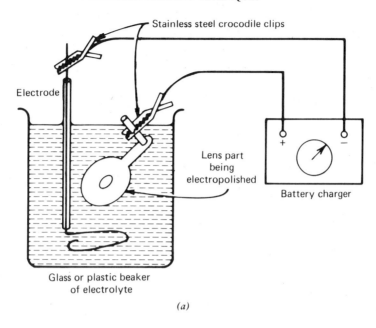

Stainless steel crocodile clips

Electrode

Lens part
being
electropolished

Battery charger

Glass or plastic beaker
of electrolyte

(*a*)

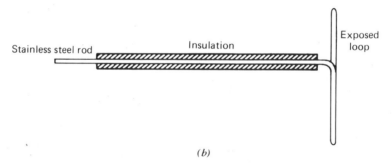

Stainless steel rod

Insulation

Exposed
loop

(*b*)

Figure 137. Electropolishing of small parts. (*a*) Apparatus. (*b*) Electrode detail.

direct flushing of the workpiece using a syringe will help ensure an even polishing effect over the whole surface. One to two minutes or even less is ample for polishing with minimum removal of metal. It is always better to electropolish for a short period, inspecting frequently, than to remove too much metal either by using too high a current or by leaving the component too long in the solution.

The recommended electrodes have been designed as current-limiting devices to minimize risk of damage to workpieces. The use of a stainless steel beaker as a cathode electrode is not recommended; it is possible to dissolve a complete lens rapidly using this arrangement.

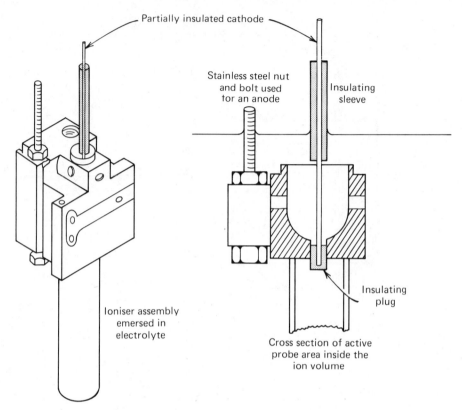

Figure 138. Electropolishing technique for internal surfaces such as ion volume assemblies.

Electropolishing the 4000 ionizer body, in particular inside the ion volume, requires a little more attention to detail. The main problems to be overcome are (1) effervescence in a confined space, which is difficult to remove and results in uneven distribution of charge and (2) localized overheating, which can cause pitting and erosion of the surface, the opposite to the desired effect.

It is essential that the ion volume and other areas that become excessively contaminated be first cleaned manually if electropolishing is to be successful. A very dirty source may require considerable effort at first, but the effort will be rewarded at future cleaning periods because the process will be so much easier. The entire ionizer body should be degreased in a suitable solvent, such as chloroform, followed by a 10-min soak with no current flowing in the electropolishing solution. This will normally soften or dissolve most organic deposits.

To remove traces of silica, most frequently found in the ion volume and on the ionizer lenses, a 10-min soak in 10% hydrofluoric acid will help.

CAUTION. **Great care must be taken when handling hydrofluoric acid.**

For electropolishing the Finnigan 4000 source, the ideal vessel to use is a plastic kitchen beaker. The source body is placed in the beaker, ion volume uppermost, as shown in Figure 138. Electrical connection is made by placing a stainless steel bolt through the heater hole extending upwards. The electropolishing solution is then added so that the whole ionizer is covered to a depth of approximately 10 mm. The stainless bolt then extends out of the solution to be connected to the anode from the power supply. The probe cathode is inserted so that the insulated end fits loosely into the solids probe inlet hole and is clamped in place. This leaves the exposed electrode roughly central in the ion volume. Approximately 2 min electropolishing (6 V, 1 to 2 A) should be adequate to achieve a reasonable degree of polishing. During electrolysis great care must be taken to flush solution continually through the ion volume to dislodge gas bubbles. Also take care to inspect the ion beam entrance holes to watch for excessive erosion.

After electropolishing, all the metal surfaces need to be deactivated by a 10-min soak in 10% nitric acid. Finally, rinse thoroughly with distilled water and then with methanol, air-dry, and reassemble the source.

Electropolishing of complex shapes such as the Finnigan 4000 ion volume is not recommended as a regular cleaning method. Users have reported that conventional cleaning methods are more effective with an electropolished ion volume and cleaning is not required as often. Electropolishing for every fourth or fifth source cleaning seems to give satisfactory results. This type of approach is probably suitable for other makes of ionizer with complex geometries, but the operator is advised to experiment on some scrap pieces of stainless steel before risking an expensive piece of the mass spectrometer. Consult with other users. Often other laboratories will have evolved techniques such as the one just described.

Plasma Ashing

Large and open-geometry ionizer and flight tube parts can be cleared of organic material by a gas plasma ashing apparatus. This consists of a chamber in which the object to be cleaned is placed. Mass spectrometer sources or slit assemblies need not be taken apart providing their construction is sufficiently open. The chamber is purged of air and then filled with a low-pressure gas. Around the chamber is a large coil to which a high-powered rf signal is applied. This generates a vigorous plasma in the gas that attacks organic deposits and removes them as gaseous oxides.

The technique is very effective on both stainless steel and ceramic parts of the relatively large and open sector instrument EI sources, but not so effective with enclosed CI sources and the smaller quadrupole sources. It does not remove silica and so is not the complete answer to source cleaning. It is not suitable for all materials. Brief experiments at Finnigan LTD's applications laboratory with molybdenum quadrupole rods were disastrous, leading to severe damage to the rod surfaces. Different gas plasmas might be more successful but no further tests were tried. The technique is useful in some cases, but it is advisable not to experiment with expensive items.

Analyzer Cleaning

Most parts of the mass spectrometer mass analyzer can be cleaned by one or more of the techniques just discussed. Some parts such as quadrupole rod assemblies and ESA plates in sector instruments are assembled to extremely fine tolerances. Cleaning of these items may be a relatively straightforward process, but advice should be sought first from the instrument manufacturer. Never dismantle any part of a mass spectrometer system without first referring to the operator's maintenance manuals, the manufacturer, or its representatives. In some cases special jigs are necessary to assemble the instrument to the required tolerances.

MULTIPLIER CLEANING AND REJUVENATION

Multipliers are extremely susceptible to contamination. The copper-beryllium box-and-grid or venetian-blind types cannot be cleaned as such, because exposure to air and water vapor reduces their efficiency. However, rejuvenating procedures are usually successful and relatively straightforward. Continuous dynode types are not so easily rejuvenated; the process is not always successful, but they can be effectively cleaned.

Continuous Dynode Multiplier Cleaning

The first step in the cleaning process for CDEM types is to ultrasonically rinse the continuous dynode assembly in a Decon solution or similar laboratory glassware cleaner. This should be followed by successive ultrasonic rinses in distilled water, acetone, and methanol. The multiplier should be thoroughly dried in a warm enclosure before reassembly into the GC/MS system.

Allow additional pumping time to clear any traces of solvent from inside the multiplier before switching on the power. Bring the operating voltage up in stages over a period of about 10 min to the maximum operating value. Do not have the ionizer switched on, or a high level of current could be produced in the multiplier and it could be destroyed. Let the multiplier sit at the high voltage for 10 to 15 min and then reduce the voltage to the normal operating level. Figure 139 shows typical before-and-after curves for a dirty multiplier.

The procedure of bringing up the voltage to a high value tends to drive dust from the surface and reduces spurious noise effects. It is worth trying

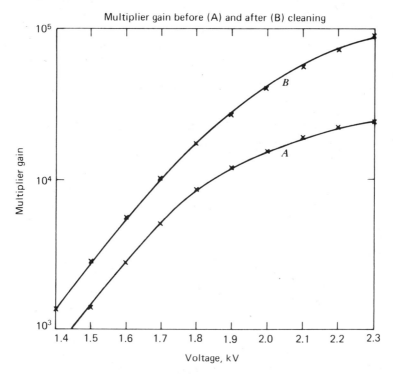

Figure 139. Continuous dynode electron multiplier (*A*) before and (*B*) after cleaning. (Courtesy of Freshwater Fisheries Laboratories, DAFFS, Pitlochry, Scotland)

on any noisy multiplier that may be suffering with loose particles on its active surfaces.

Box and Grid or Venetian Blind Rejuvenation

To rejuvenate a multiplier with copper-beryllium surfaces, remove it from the mass spectrometer and immerse it in high-quality ethyl acetate. Agitate the liquid by using an ultrasonic cleaner for 10 to 20 s. Rinse it thoroughly in the ultrasonic cleaner, first with acetone and then with methanol. Dry the multiplier quickly with tissue and then wrap it in cleaned aluminum foil. (Aluminum foil usually has a fine coating of lubricant from the rolling process. Remove it by first baking the foil in an oven. Take care not to touch the heater element with the foil by accidentally poking it through the heater's protective mesh.)

The wrapped multiplier should be baked at about 325°C for 30 min. A gas chromatograph oven is ideal for this. After baking, cool and reassemble the multiplier as quickly as possible. Try to avoid carrying out work on copper-beryllium multipliers on humid days or in a damp environment. Pump the vacuum system down as quickly as possible after cleaning.

Continuous Dynode Multiplier Rejuvenation

To rejuvenate continuous dynode electron multipliers, place the multiplier assembly in a beaker containing a warm solution of 10% hydrogen peroxide, 10% formic acid, and 80% water.

CAUTION: **This solution is extremely toxic and produces noxious gases. Carry out this procedure under a well-ventilated fume hood. Do not let the liquid temperature exceed 80°C, or the reaction may become very violent.**

After 5 min, rinse the multiplier by running warm water into the beaker for at least 5 min to remove all traces of the rejuvenating solution. Lift out the multiplier while the water is running over it to prevent dirt pickup from the water surface. Blow the multiplier dry with a clean gas line such as a nitrogen line. Complete the drying process by wrapping the multiplier loosely in tissue and warming it to about 80°C in an oven for about 15 min. Reassemble the multiplier into the mass spectrometer, and pump the system down. Apply power to the multiplier slowly, as described above, only after thorough pumping.

Multiplier Gain Checks

Multipliers lose their signal gain with use (Fig. 140). Cleaning and rejuvenating processes help restore it, but rarely is it completely recovered, and in time the multiplier must be replaced. Maximum gain for a new multiplier is 10^6 to 10^7, and for some is even better. Conditions such as the multiplier voltage affect the signal gain, and operation in the 10^5 to 10^6 region is normal. As the maximum available gain falls, the drive voltage is increased to maintain the operating gain. There comes a point in time when this can no longer be done because of increasing multiplier noise. On the other hand, a low-gain multiplier may still be useful providing the noise level is low. Systems have been known to perform really well with a multiplier operating at less than 10^4 gain because the noise level was so low. Signal-to-noise ratio, not absolute gain, is the most important parameter.

An absolute measurement of the multiplier gain is consequently of doubtful use. However, plots of gain against voltage and their change with time are useful indicators. On some instruments the gain cannot be measured, but others have Faraday cups or x-ray shields that can be used as Faraday plates. The signal before the multiplier can be measured on the Faraday plate collector when it is switched off. The Faraday cup is then moved out of the ion path, or in the case of x-ray shields the multiplier is switched on and the signal is again measured. The ratio of the two signals gives the gain. The procedure is usually described in some detail in the operator's manuals. Unfortunately, to see a signal reasonably on most GC/MS instruments using just a Faraday cup or plate requires a strong signal. The system was not designed to be used this way. Switching on the multiplier with this artificially high ion signal can damage it, due to ion-electron cur-

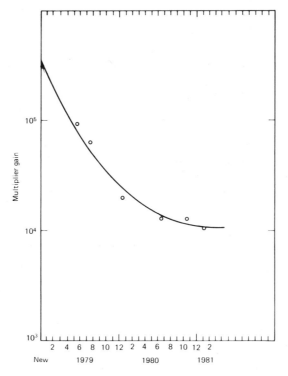

Figure 140. Continuous dynode electron multiplier aging characteristic over a 24-month period. Gain at 2-kV decrease over time. (Courtesy of Freshwater Fisheries Laboratories, DAFFS, Pitlochry, Scotland)

rent burns on the active surfaces. Furthermore, the current down the multiplier may saturate and give a reduced gain reading. The test is open to doubt and in some cases is likely to damage the instrument.

A better approach is to observe a reasonably strong but normal signal with with the multiplier switched on. Switch off the multiplier, and try to observe the signal using the Faraday cup, increasing the electronic amplification by a factor of, say, 10^4 or even 10^5, if possible. If the signal is observable, the gain can be calculated. If not, then the multiplier gain is better then 10^4 or whatever the amplifier-display gain change was. This "better than" figure is probably of greater value in checking that everything is satisfactory.

VACUUM SYSTEMS CARE

There are two schools of thought with regard to the maintenance of mass spectrometer vacuum systems. The first holds the view that one should leave well enough alone and disturb the pumping system only if it is absolutely necessary.

The second argues that regular overhauls are the only way to keep an instrument in peak performance. Up to a point, both arguments are valid.

There is a lot to be said for leaving the pumping system alone. First of all, it is the easiest option. Also, pumping systems tend to be self-cleaning if sample rates are low. Unfortunately, few laboratories have the luxury of allowing the system to recover by slowly pumping away with all inlet ports sealed off. This would be downtime as far as samples are concerned and in most situations must be minimized. Maintenance should therefore be anticipated, and it should be part of the instrument operating schedule. The best approach is to decide on a maximum time interval between overhauls, shortening the interval if necessary or if a suitable opportunity arises. This approach should give the best operating conditions with minimum downtime for maintenance.

Whichever approach is adopted, the vacuum system should be checked regularly so that any problems can be quickly detected. The operator should know the normal indicated vacuum pressures for all the operating procedures used. Be aware of the background spectrum. With computerized routines for subtraction and enhancement it is easy to forget that it is there. Background peaks can tell you a lot about the condition of the system.

Rotary Pumps

Check the level of rotary pump oils every week, and top up as necessary; a low oil level causes severe backstreaming and can put diffusion pumps at risk. It is a good idea to keep the pump standing in a tray to catch leaking oil. Monitor leaks, and order replacement bearing seals and gaskets in time for the next maintenance period. In fact, once you have decided to strip a pump, it is a good idea to give it the full treatment. Most pump manufacturers sell an overhaul kit of parts. Fitting these is usually straightforward, though often a messy job. Mechanical wear is inevitable, and you must expect to do this job every two or three years.

Direct-drive pumps operate at higher speeds than belt-driven units and are more prone to leaks at their shaft seals. With some makes of pump it helps to stand the pump on its end with the motor uppermost. Check with the manufacturer on whether the pump can be operated in this position. Make sure that the oil filter hole and drain plugs are secure, and, finally and most important of all, make sure the pump is supported and will not fall over.

Belt-driven pumps operate at slower pump speeds and do not seem to suffer quite so much from leaking shaft seals. The critical item with these pumps is the drive belt, and it is a good idea to carry a spare. Since the belts are usually enclosed in a protective guard, failure is difficult to anticipate. Even when the drive belt is visible it is not always obvious that it is worn out until it actually breaks. Check the pumps every day and get used to the sound they make. The vacuum system pressure may hold for a while on the other pumps, but when a drive belt goes or a motor fails the sound of the instrument will definitely change.

Rotary pump oil should be changed about every six months. With heavy use, it should be changed every three months. Treat used pump oil with care. It will contain concentrated residues from samples run on the GC/MS system, and these may be harmful compounds. Follow approved procedures for disposal of contaminated oil.

Pump oils contribute significantly to the background spectra, and it is certainly worth trying a number of different types or taking advice from other users of the same type of instrument operating in the same circumstances. Foreline traps help considerably, but they too do not last indefinitely. The trap should be replenished or regenerated every time the pump oil is changed, and more often if background levels are critical.

Gas ballasting or pump purging is a useful way of removing water and solvent from rotary pump oil. If the gas ballast valve is opened, there is a high throughput of air. The design of the pump usually restricts the leakage of the air into the vacuum chamber, but the pumping speed is reduced and some backstreaming of pump oil will inevitably occur. Gas throughput on most GC/MS systems is high, and consequently it is not usually necessary to ballast as a matter of routine. It is better to hold this in reserve, only using it when needed. On some systems the gas ballast valve can be opened fully without a big rise in pressure; on others the valve must be only partially opened. It is a sensible precaution to switch off the mass spectrometer electronics first.

Rotary pump oils tend to creep up the backing tubes. In addition, material can condense on these tubes and will be pumped away only slowly. This accumulated material is a source of contamination, and after a couple of years the tubes should be replaced. The tubes also stiffen, crack or perish with age, and it is a good idea to replace them every year or two. New tubes tend to outgas a lot at first, and the background level should be monitored carefully. It should come down to a reasonable level after a few days of pumping.

Rotary pumps transmit a lot of vibration to the mass spectrometer and can be a considerable source of microphonic noise in the electronics system. Electron multipliers are particularly susceptible to vibration pickup. Take care when servicing rotary pumps that the flexible tubes are not constrained or resting against part of the mass spectrometer chassis. Standing the pumps on a cork mat is a good way to reduce vibration and also their general noise level.

Mechanical pumps generate a lot of heat, both from their motors and in their pumping stages due to friction. adequate ventilation is most important. The motor usually has an integral fan, and an unimpeded air flow is all that is required. The pumps usually stand on the floor, and the motor fan tends to suck in a lot of dust and fluff. Make sure the pump motors do not become clogged. Out of sight should not be out of mind.

Diffusion Pumps

Under normal operating conditions, there is little maintenance to be done on diffusion pumps. Certain designs have a tendency to lose oil, and the oil levels should be checked once a year and topped up as required. Their pumping action tends to be self-cleaning, so a complete overhaul is necessary only after a few years of operation. High throughputs of air and vacuum accidents reduce the operational lifetime. The jets become dirty with a sticky tar that is formed from cracked oil and contaminants. The main charge of fluid becomes discolored, sometimes with a lot of particulate matter in it. In these circumstances, the oil should be drained off and the pumps cleaned. It is easier to remove the oil while it is still warm.

Cleaning the pump involves washing it with solvents to remove the last traces of old oil and then scrubbing it to remove the varnishlike deposits.

NOTE: **The pump should be allowed to cool before washing with solvent, or the fire hazard due to solvent vapor may be increased. Some solvents, such as trichlorethylene, are not flammable but give off harmful vapors and should be handled with care.**

An effective and safe way of cleaning diffusion pump jet assemblies is to boil them in soapy water and then rinse them in boiling water. The lacquer deposits can then be removed by shot blasting. This is a specialist service and is often available locally from high-quality metal finishers. It takes only a few minutes to get a very high polish. The blasting powder should be glass beads, since they give a smooth finish. Pumps polished in this way can achieve remarkably good vacuum pressures in a very short time and tend to stay cleaner for longer. Some sand-blast machines leave an unsuitably rough and very porous surface.

The operation of a diffusion pump relies on a cool outer surface to condense the pumping fluid. Pumps are usually fitted with protective devices to prevent drastic overheating, but even slight overheating can cause an increase in oil backstreaming. Check regularly to make sure fans are not blocked or siezed on air-cooled models and confirm adequate water flow in water-cooled systems. In areas where the water is very hard, furring up of the pipes is a danger, and a closed recirculating system with a chiller is recommended. The water consumption of a typical mass spectrometer system can be quite considerable, and a recirculator may be justified on running costs alone. It should be filled with soft water with an added anticorrosion, antigrowth mixture. Commercial antifreeze solutions are usually suitable. Significant algal growth can otherwise occur, and very soft waters without additives can eat away the cooling pipes.

Another cause of overheating is a clogged water filter. The filter element should be changed at regular intervals. The time scale depends very much on the individual location. Change the filter at least once a year, more often if dirt buildup is a problem. Some laboratories find that a reusable coarse prefilter that is cleaned every month or two is a good idea, the very fine main filter needing only an annual change. A bypass valve arrangement can be fitted to the coarse filter to allow it to be cleaned without shutting down the vacuum system.

Although overheating is the most common diffusion pump problem, over-cooling can also occur. If the water supply is very cold, the entire pump assembly will cool and the oil will boil inefficiently, reducing the pump's level of performance. Check the manufacturer's specifications for the pump, so that conditions can be set for optimum results. This will pay dividends in terms of rapid pumpdowns and low background levels.

Turbomolecular Pumps

Turbomolecular pumps require almost no maintenance. The word "almost" is very significant. When turbo pumps fail, the result can be spectacular. A disintegrating turbine blade or bearing rapidly reduces the rest of the pump to a pile of scrap metal, which, unfortunately does not just fall to the bottom of the pump but is usually accompanied by a rapid ingress of gas due to the loss of pumping action, and debris accumulates in all parts of the mass spectrometer. Slits, flight tubes, quadrupoles, multipliers, filaments, and what have you can be severely damaged. Hence, it is vital that the manufacturer's recommended maintenance be carried out. This usually involves changing a small amount of special oil in the bearings every 6 or 12 months. Other designs require new bearings after a certain number of hours of operation. *Do not delay these jobs.*

Some turbo pumps require a cooling air or water supply; this is primarily to reduce the amount of bearing oil backstreaming into the vacuum system. As with diffusion pump cooling, follow the manufacturer's instructions. Monitor signal noise levels on GC/MS systems with turbo pumps. An increase in high-frequency noise on signal traces can be indicative of additional vibration from the pump. An increase in vibration is usually accompanied by a rise in the audible noise of the system. These sounds may just be caused by badly positioned backing pump tubes, incorrectly adjusted mounting brackets, or even loose components within the mass spectrometer. All cases are worthy of investigation, not least because they may indicate the early stages of a turbo pump failure.

Vacuum Seals

Vacuum seals need not be replaced as a matter of routine. That is, there is no fixed time scale for when a given O-ring or gasket should be replaced. Replace inaccesssible seals only when that part of the vacuum system is being serviced. Other seals, such as those on probes, will show appreciable wear in use and should be replaced as necessary. It is better to replace a seal if there is some doubt about the quality of the existing one. Never try to get extra life out of a seal by overtightening fittings. This advice is even more relevant to gas-line fittings, as the tubes are easily damaged. Always have a set of O-rings, gaskets, and ferrules available so that changes can be made as required without delay.

Vacuum leaks usually occur at pipe and flange fittings, often just after a seal has been replaced. Do not tighten fittings indiscriminately if the system has a leak, or additional leaks may be created due to overtightening. Try to locate the source of the leak and adjust only that fitting, replacing the seal if necessary.

PERIPHERAL EQUIPMENT CARE

Nearly all GC/MS instruments contain OEM modules, units made by another company and built into the system by the GC/MS manufacturer. Even if all the units carry the same company name, parts may have been made in a number of different divisions. These units normally have their own set of manuals, which are rarely used by the GC/MS operators. Their main use is to assist service personnel if faults occur, but they often contain useful regular maintenance checks and schedules. It is worth identifying these tasks, as regular maintenance can be vital to continuous reliable operation.

Typical of this type of maintenance is the care of the computer system disk drives. The recording heads float on an extremely thin layer of air above the disk. An adequate supply of very clean air is provided by a blower arrangement with a very fine filter. The filters are good for only a few months of continuous operation and should then be replaced. Minimize the amount of dust around a data system. Remove disks only when necessary, and store them in a clean environment. Even with the greatest of care, after a year or two disks can become dirty and even mechanically distorted. There are companies that will check and clean disks and other recording media. In some cases data are guaranteed, but as a precaution it is better to copy important information to another disk or tape before using the service.

Disk recording head alignment is another critical parameter that can change with time. Disks that were once readable can become useless. Data recorded on a misaligned drive cannot be read elsewhere or on the original drive if the disk is realigned. It is important to have drives checked at least once a year to ensure reliable alignment. The problem is even more acute in systems with more than one disk drive. Despite what the manufacturers say about interchangeability, it is good practice to assign certain disks to one drive, others to the second drive, and so on. If disks must be used in more than one drive, try to restrict write operations to one of them. The very nature of the computer system is such that one tiny error can often make the entire disk unreadable. A similar situation occurs with tape units. These are not usually quite so critical on alignment, but they are subject to more wear and tear due to contact with the read/write heads. Always keep the tape units clean and store tapes in their cases in a clean place.

Tapes and disks have significant warmup characteristics. Try to keep the ones that are used regularly at about the same temperature as the data system— in other words, in the same room. Allow at least 5 min warmup time over and above any delay set by the system if you are going to write to the medium. Longer warmup (or cooldown) times will be necessary if the tape or disk was stored away from the system. Read-only operations can usually be performed as soon as the system is ready, but some errors may be reported if the media temperature has not stabilized. Warmup times are frustrating and inefficient in terms of instrument use. Delays can be minimized by careful management of the recording media.

Even on instruments without data systems, recording devices are the main peripheral area requiring regular maintenance. Ultraviolet recorder lamps, for instance, have to be replaced fairly often, as their intensity soon fades. The lamps are nearly always made from quartz glass and operate at high temperatures. Try not to touch the glass, as finger oils and acids can etch the surface, reducing the light transmission or even causing the lamp to burst. If the glass is inadvertently touched, wash off the fingerprints with methanol and polish the glass dry with a soft tissue. Always wear safety glasses for protection when handling these lamps.

The lamp mounting brackets and electrical contacts are exposed to excessive heat from the lamp and can oxidize. Occasional cleaning of the contacts can often improve the starting, lamp striking, characteristics. Performance and operational life can be enchanced by setting the power level correctly. Do not change a bulb without following the manufacturer's setting-up procedure.

Strip-chart recorders can be a great source of irritation if they are not kept clean, correctly adjusted, and lubricated. Pens, in particular, are a regular cause of trouble. Storing the pens in a humid container, such as a jar with a piece of damp sponge in it, when they are not in use seems to be one answer to poor starting as this prevents the nib from drying out. An occasional cleaning and refilling with new ink helps to prevent starting and blotting problems. Keeping the recorder clean and free of ink spills is the only way to ensure that the charts are not inadvertently spoiled. Sticking slides, dirty slidewires, and incorrect mechanical or electrical damping adjustments (called "gain" on some recorders) can cause noisy traces. Increasing the damping (reducing the recorder servo-drive amplifier gain) too much can be just as serious a fault. The recorder response becomes sluggish, and the recorder will not record the amplitude of peaks accurately.

Recorder sensitivity ("span" on some units) should be checked regularly. The different recorder range settings may not be exactly as indicated on the front panel; an error of a few percent may be quite critical in work involving a wide range of sample concentrations. Changing a sensitivity setting can introduce errors if the selector switch is not calibrated. this applies to the mass spectrometer controls as well as to the recorder. It is wise to have the system calibrated at least once a year, more often if very precise work is required that may involve changing recorder channels or settings. All the safeguards inherent in using internal standards will be lost if the working ranges of three or four recorder settings are not calibrated. If the standard and the unknown are recorded on the same channel with the same settings, then there is no problem, but with multiple ion detection systems, this is rarely the case and regular calibration is necessary. Calibration need not involve sending the equipment away. Small dc reference sources designed for this purpose can be purchased from chart-recorder manufacturers or even made using a battery and a selection of resistors. Relative performance of recorder channels can be checked using these reference voltage supplies or by measuring the same background peak with all channels. Up and down one setting can usually be measured with sufficient accuracy to check their relative sensitivity.

PART THREE

TROUBLESHOOTING FAULTS ON GC/MS SYSTEMS

CHAPTER ELEVEN

The GC/MS Instrument Under Fault Conditions

It would be unrealistic to expect trouble-free operation of a system as complex as a GC/MS instrument. From time to time faults will occur and must be rectified. The first person to notice the fault is usually the operator, and he or she has to decide what to do. The operator is faced with trying to solve the problem alone or calling for help. The help may come from within the organization or from the GC/MS manufacturer or its agents.

Consider the following telephone conversation between a GC/MS operator and a service engineer.

User	My system is not working.
Engineer	What exactly is wrong?
User	No sensitivity.
Engineer	Can you see any peaks?
User	Oh sure, but sensitivity is terrible.
Engineer	What sort of levels can you see? Have you run any tests?
User	Not sure, really. It was okay last week.
Engineer	What levels could you see last week?
User	Oh, much better than this week.

The conversation then loops back, and so it goes on. In the end, the user and the engineer were both frustrated and a visit was arranged. The engineer's journey of 100 miles was a long way to travel to change a septum, clean the separator, turn up the electron energy, switch off the X100 filter, sort out the samples from the blanks, or whatever. Most service engineers have had to do these things at one time or another. The user could have saved everyone considerable frustration, expense and delay by defining the problem a little more clearly.

DEFINING THE PROBLEM

First of all, is there really a problem? Check the obvious things first. This can save a lot of embarrassment later. Second, investigate the apparent problem a little before reaching for the telephone even if you do have a five-star service contract. You may be able to fix it yourself or at least define the fault so that the engineer can send or bring the correct replacement parts. Obvious hardware faults like burned-out oven fan motors or gas-line solenoids are easy to diagnose, but many electronic hardware problems manifest themselves in strange ways. A basic knowledge of the instrument's electronic modules, especially power supplies, and their interrelationships is invaluable in these situations.

Intermittent faults are harder to define. Try to note the conditions leading to a particular problem and check for any recurring pattern. Does the fault occur at the same time of the day with the same operating conditions, or after a particular sequence of events? Remember, faults always disappear the moment the service engineer arrives, so some evidence is necessary to keep him around long enough for the fault to reappear. Software faults are a particular case of intermittent problems. Try to note carefully the sequence of operations leading to the fault. Is it reproducible? Remember that even experienced operators can make mistakes. Be confident in your ability and your memory of how to perform a particular operation, but do not get overconfident. If things do not work as you think they should, check out the procedure in the operator's handbook or with an experienced colleague. Many apparently severe data system faults stem from operator "finger trouble."

Chromatography and sensitivity problems are probably the hardest to define. Furthermore, the average service engineer will not have a lot of experience in this area. Most service work is done on broken instruments and the service person has little opportunity to develop GC/MS operator skills. On the other hand, this is the area in which you the operator have the most experience, and a lot of self-help is possible if the problem is tackled logically. Reference to previous results (record keeping) and running standards (system checks) are the obvious way to tackle this type of fault.

LOCATING THE PROBLEM

Having decided there really is a fault, try to narrow the field to isolate the problem area. Investigate the fault and note anything unusual even if it does not seem to have anything to do with the problem. For example, a customer once reported very bad noise and peak instability. The fault appeared to be an unstable power supply board. It was also reported that a new solenoid sent as a replacement was defective and probably also leaking, since the vacuum pressures fluctuated badly. Investigation of these "unrelated" faults revealed a burned-out cable to the solenoid, which was causing severe ground current loops. This caused very unstable power supplies to the mass spectrometer and the vacuum controller. Fixing the cable fixed all the problems.

One clue to solving this problem was noticing that although the display wobbled about, the background peaks did not go up with the rise in the pressure display. Switching the solenoid out of circuit by selecting a particular operating mode cleared the display noise problem as well as appearing to fix the vacuum monitor. The problem therefore had to do with the electronics and not the vacuum as it first appeared. The key to solving the problem was *observation* and *valid deduction* confirmed by a number of tests.

The instrument log book is a useful source of information when trying to pinpoint intermittent faults. Patterns or common features can be identified. Wherever possible, procedures should be repeated to establish consistency. In cases of sensitivity and chromatography faults, work can be repeated so that sample-related phenomena can be identified. Column and system performance can be reliably checked against previously known standards. GC/MS instruments are not designed to go wrong, but even so, considering their complexity, it is surprising that there are so few faults. Many faults are brought about by disturbing the instrument's status quo. When faults occur, always think back and try to determine whether you or anyone else has disturbed or altered any part of the instrument. It is amazing how many drastic electronic failures occur after a source cleaning that seem to fix themselves when the source is again removed and then replaced in the instrument!

TROUBLESHOOTING THE ELECTRONICS SYSTEM

Problems with particular parts of the GC/MS instrument are discussed in some detail in the following pages. However, most parts of the instrument contain a significant amount of electronics. Consequently, some general comments and troubleshooting hints with respect to the electronic and electrical systems are appropriate.

Electrically, a GC/MS instrument is a complex arrangement of power supplies, various circuits, and controllers with a host of interconnections. This system can fail in a number of different ways. A permanent fault may occur, with obvious consequences. On the other hand, the result of a malfunction may be quite unexpected; all that is certain is that something is wrong. Add to this the possibility of intermittent faults, and troubleshooting the electrical systems becomes quite complex and is beyond the scope of the average operator. However, there are certain types of fault, that are common.

Faults do not occur frequently, but when they do, they often fall into recognizable categories. While the operator may not have the skills, equipment, or parts to fix the instrument, at least if the problem can be defined then a successful repair is more likely in a short time. A large proportion of service calls on GC/MS instruments are to fix bad connections, cables, or defective power supplies. A basic knowledge of the GC/MS electrical systems is all that is usually necessary to identify the problem area.

Connectors and Cables

Faults on connectors and cables are the major cause of intermittent problems. In this category also comes printed circuit board edge connectors and switch contacts. When a fault occurs, try to locate the problem area from the symptoms. Never waggle or remake connections indiscriminately. More faults are caused than are fixed by this approach. However, if the choice is narrowed to two or three cables or connectors, then wiggling these cables and remaking or cleaning contacts is valid.

Sometimes, the problem is noise or pickup. This can occur if a cable is lying next to a heavy main-line power cable or across a rotary pump motor. The relative position of some cables can be quite important. For sensitive signals, screened cables are used. The screening may be ineffective because of poor contacts at the connector, due to dirty or oxidized pins or even loose shield assemblies. On some types of lead, especially coaxial cables, the screen is an outer braid that is crimped or clamped to the connector. Any strain or stretch of the cable tends to pull the screen clear, giving rise to pickup problems.

Oxidation of contacts, especially on power supply connectors, switches, and relays can occur either due to their current load or because of their location. Oven heater and interface heater connectors, for instance, are prone to oxidation problems. When the problem is located, clean the contacts with a piece of very fine emery paper or a commercial contact cleaner. Always make sure the unit is switched off and disconnected before touching the contacts.

Similar problems occur with rotary switches, which are also prone to considerable wear during their lifetime. A cleaning with a commercial switch cleaner containing a switch lubricant is a good idea. This can be obtained in spray cans that are generally provided with a spray tube to reach those awkward corners. Potentiometers are also subject to wear problems but are usually sealed units and the only recourse is replace them.

Modern cable connector assembly methods bring with them their own problems. Crimp-type connections are often made incorrectly, the metal being crimped or bent around the insulation instead of bare wire. In this case, contact may just be made, but it is very unreliable and prone to intermittent faults. Always look at suspect cables carefully. Some types, notably ribbon cables, are supposed to be assembled in this manner; the contacts pierce the insulation to mate with the wires inside. Although the cables are flexible, any strain on the connector header tends to break the contacts or even causes shorts between adjacent pins.

Printed circuit board edge connectors are usually gold plated but are still liable to a buildup of material that prevents good electrical contact. Clean the contacts by rubbing with an ordinary pencil eraser. Take care not to get pieces of rubber inside the edge connector; brush them clear and then wipe over the contacts with a tissue soaked in dichloromethane. As with the eraser, take care not to leave pieces of tissue on the connector.

A final word on the subject of connectors: Never assume that the pins mate just because you can see them inside the connector shells and the two halves go together. It is not at all unusual for pins to push back as the two halves come together so that contact is not made. Once again, close inspection is required.

Power Supplies

A large proportion of the GC/MS instrument's electronic system consists of power supplies of one sort or another. The stability of these is fundamental to satisfactory performance of all parts of the system. Complete failure of a power supply voltage is usually obvious. Investigation of the power supply test points is one of the first electronics troubleshooting tasks, no matter how simple or complex the electronics equipment. In fact, checking the power supplies may be the only test that the (electronically) unskilled operator may perform. If a power supply shows an erroneous voltage, the fault may lie elsewhere in the system, with other circuits causing an overload on the supply. It may be possible to unplug some parts of the instrument to reduce the power supply loading, but this should only be done after advice from a qualified engineer.

Some power supplies are potentially lethal, capable of giving high currents at high voltages. Investigate test points only if you are confident in what you are doing and have a suitable test meter with adequate test lead insulation. Never work on an instrument alone, in case of electrical shock. Your assistant should know what to do if you are subjected to a severe electrical shock. He or she must know where to switch off the power and how to administer appropriate first aid. This may sound melodramatic, but you may not get a second chance if an accident occurs.

To check the test points, you need to know where they are and what to expect. The instrument service manual will give details, and practical training can be obtained at a customer training course or on site during a service engineer visit. Many of the power supplies are protected by fuses. If the supply is completely dead, check the fuses. Once again, familiarize yourself with their location before failures occur and make sure you have the necessary spares. Never change a fuse for one of a different rating. If the new fuse blows immediately, you have a fault; do not keep on trying fuses.

Apart from the failure modes described above, power supplies are prone to problems of noise. This can take several forms. Low-frequency noise or mainline hum is associated with the main power supply and will be at the same or twice the line frequency. In the United Kingdom these will be at 50 and 100 Hz. In the United States the frequencies will at 60 and 120 Hz. Most countries adopt one of these two standards. Noise of this sort on a power supply voltage may appear as ripple on the mass spectrometer signals. The fault may be due to poor power supply regulation, poor connections, overheating, bad grounding or screening, and ground loops.

Ground loops are caused by power supplies having alternative return paths for current flow. The resultant loop can act like a shorted transformer turn in which a current can be induced by pickup from the main power supply. All would be well if it were not for the fact that the lead impedances are not negligible and small voltages, that is signals, are produced. Instruments are not generally designed with ground-loop problems, but they can occur if additional equipment or grounds are attached to the instrument. For instance, metal waterpipes and gas lines can cause this type of problem. Minimize the effect by ensuring that the main bonding between the instrument and ground is in good condition. The instrument is often grounded to water pipes or conduit systems, which may be perfectly adequate from a safety point of view but can be a relatively high impedance for small signals. Check the earth if mains ripple problems occur, and clean up or tighten the contacts to ensure a good quality ground. Make sure the various parts of the instrument are adequately grounded. Sometimes a heavy duty braided lead between consoles and manifolds reduces mains ripple.

Line-related noise at higher frequencies can sometimes be seen. This appears as a stationary or nearly stationary ripple on signal tracks and power supplies when the oscilloscope displays are triggered from the power supply line. Sometimes, under fluorescent light, which produces a slight stroboscopic effect at the line frequency, the normal spectral display will show slowly moving or even stationary glitches. This type of line noise is nearly always associated with electrical motors and may be due to pickup from the mechanical pumps or even the laboratory refrigerator compressor. Repositioning of the motor or cables can sometimes improve matters. Other types of higher frequency noise appears as "grass" on the traces. This may be due to poor cable screening and bad contacts on signal tracks. In the case of the displayed spectrum it may be due to an aging multiplier rather than the power supply itself.

One should bear in mind also that the mains power or line voltage is not fixed and may fluctuate considerably. There may be quite significant changes at certain times of the day and certain times of the year. If you experience vague instrument problems, it is always worth checking the main line voltage. Too low a voltage and the instrument power supplies will go out of regulation. Too high and the system will overheat. Most are designed to accommodate a small fluctuation and can be set to a small band within quite a wide range. If the line voltage in your area tends to vary a lot, it may be worthwhile fitting a power line stabilizer to the critical parts of the instrument. The manufacturer's service department or their representative can usually advise on this.

Electronic Malfunctions

If an electronic fault is suspected and the power supplies appear good, it will be necessary to troubleshoot the electronics system. In most cases, this is where the operator calls for help. In some cases, the problem is obvious or the operator skilled enough to tackle the faultfinding him- or herself. Most service or maintenance manuals will give basic electronics descriptions with test-point data and, except for very high voltage circuits, the signals can be followed through using an oscilloscope. In some cases the instrument's own display monitor can be configured for test purposes. On some of the automated GC/MS systems, test programs are available that lead the operator through a series of checks, the program prompting the operator via the display terminal and the data system configuring the instrument into different test modes. Figure 141 shows a complete troubleshooting flow diagram for a Finnigan-MAT 1020/OWA system. The right-hand part of the procedure calls up tests to check the analog electronics. The left-hand section refers to data-system testing. The aim of any test procedure, whether manual or fully automated, is to progressively narrow down the problem area.

Where possible, independent checks of different parts of the system should be performed. This can often be achieved by operating the instrument in different standard modes or even by configuring the electronics by changing some of the connectors. Input test signals can often be generated using a standard test meter. Setting the meter for ohms readings causes an internal battery to be connected at its terminals. Resistors should be connected in series with the leads to limit the current flow. There are a number of tricks of the trade using such arrangements, and these can often be picked up from an experienced service engineer or at a customer training course.

CALLING IN THE EXPERTS

There comes a point in any troubleshooting situation where additional help may be required. The source of help will depend on a number of factors: the nature of the problem, whether the instrument is still under warranty, whether the laboratory has a service contract, and what help is available locally.

In-House Experts

Many large organizations have their own electronics section with highly skilled technically qualified personnel. It would seem sensible to call upon their services if an electronic fault should occur. Unfortunately, these people are rarely familiar with the GC/MS system, either as a scientific instrument or as a group of electronic circuits. It is very difficult for them to relate the symptoms of a fault to parts of the instrument. Although their understanding of the circuit diagrams may be very good, it is difficult for them to comprehend the whole electronics system without training and experience on the instrument. Often a search in the schematics manual compounds the problem, as the helper may have some doubts about the validity of a given circuit design. It is always possible to improve a circuit or achieve the working function in a different way, but this is no time for redesign. The given circuit did work and now it does not. It is not fair to

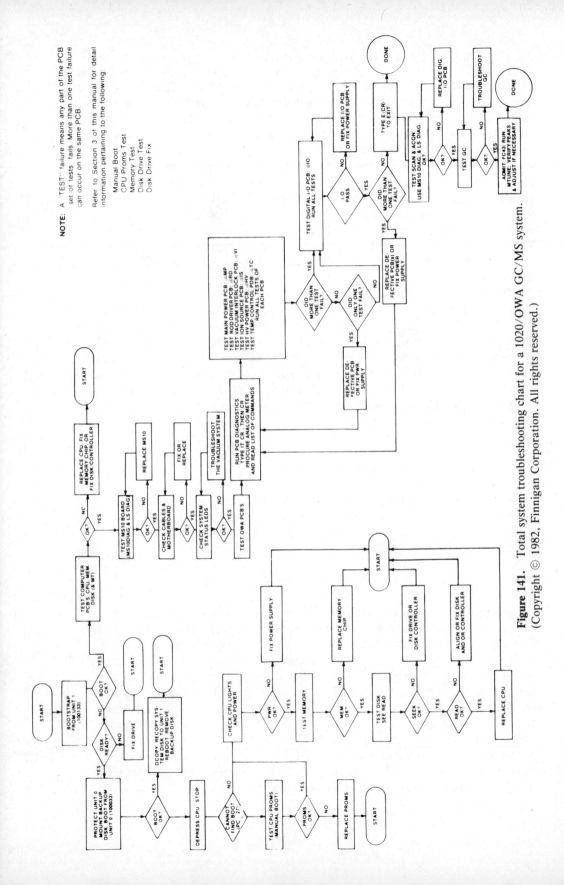

Figure 141. Total system troubleshooting chart for a 1020/OWA GC/MS system.

throw the unfamiliar expert into this situation and is unlikely to solve the problem. You may end up with beautiful electronics but a GC/MS instrument that still does not work.

A more successful approach is for the operator to define the problem in more detail, if necessary with telephone support from the manufacturer. The skilled local expert can then be called in to make the more detailed checks and do the repairs once the problem area has been narrowed down. It is often possible to pinpoint a problem in this way to a small group of components, but only the skilled electronics expert with the right test equipment and tools can make the final on-the-spot diagnosis and effect the necessary repairs. This approach also has advantages in that it reduced the amount of time the local expert has to spend on the problem and also informs the manufacturer's service department that action may be required by them.

Professional Help

Service Contracts.   There are many different types of service contracts offered by the manufacturers, and some include emergency service work. The basic contract usually covers one or more visits to carry out routine maintenance tasks. This can be extended to cover a few emergency visits or even a full, no-limit, minimum-time-response service. Where the contract covers only a limited number of calls, be wary of calling out the engineer at the first sign of trouble. It does not take many incidents to use up the allotted service time. Certainly, call for guidance on a problem, and follow this with a request for help if you need it.

With a full contract that covers all maintenance and service, there is a tendency to shout for help immediately when a problem occurs. Once again, it is a good idea to telephone for advice and to arrange for a service call if necessary, but you should still be prepared to do some of the troubleshooting work. This helps to define the problem clearly, so that the right spares are obtained and the engineer arrives well briefed. Even the best contracts only guarantee a service response time with spare units available for replacement or hire. They do not promise to fix the instrument instantly. It is better for the engineer to arrive with the necessary bits to repair the instrument than to get there more quickly but unprepared for the problem at hand.

Telephone Service.   Before telephoning for help, make sure you can clearly describe the problem. If there is doubt as to its exact nature, be specific about the symptoms. In other words, have the facts ready. This means gathering together any notes about the problem, the instrument log for the time period leading up to when the fault was noted, and all the details of any tests carried out. These may include test point data and the current instrument parameters in the various operating modes.

Have the instrument service and maintenance manuals to hand. While these may be outside the normal operator's area of understanding, the service engineer may wish to refer to a particular section, diagram, or photograph during your discussion as an aid to troubleshooting. This is especially true when trying to sort out vacuum or interconnection problems. For those who are happy to investigate electronic faults in detail, the service books provide the necessary basis for discussion and usually give details of test point locations.

Telephone from the GC/MS laboratory, not another room, so that any checks and tests requested can be quickly carried out. Having a telephone near the instrument allows the service engineer to talk you through a procedure.

Always remember that the person at the other end of the telephone is unfamiliar with your problem. He or she may know your type of instrument very well but be unsure of the exact configuration in your laboratory. It may be that he or she is not really service oriented and may have to seek advice on your behalf from someone else. Allow them time to understand your setup and the nature of your problem. Be patient when they ask seemingly obvious questions or ask you to repeat certain measurements. Double-check a measurement if asked to; it could save a lot of embarrassment later on, and it does give the service supporter a chance to think through your problem.

If after receiving advice you try again to resolve the problem yourself, be sure to let the service engineer know the outcome. Obviously, if the difficulties persist, you will continue the dialogue, but if you resolve the problem or even decide to live with it for the time being, then let the company know. They may be holding back spares, delaying a job or another call, waiting for word from you. Calling back with a quick okay not only helps other customers but also builds up the service department's experience as they get feedback on a variety of problems. This will enhance your credibility and should help you get good service in the future.

Service Visits.   Never be afraid to call a halt if you are being asked to dig too deeply into the instrument. Only take things as far as you feel you are confident to handle. Be clear in your instructions to the service engineer at this point. If a service call is required it will be necessary to have an authorized order or contract number. If this has to be sanctioned, let the service engineer know you will call back with the details. It may be that a visit will be arranged or at least planned before the order is received, but you should not ask for one without providing the necessary authorization.

Be available to welcome and help the service engineer. Warn the site security that someone is expected and arrange parking near the laboratory or, if that is not possible, arrange for someone to help carry tools and equipment. The engineer may have had a long journey to get to your laboratory. Do not expect someone who got up at six and has driven for five hours so that he reaches you at a reasonable time, to be fresh and raring to go. No matter how angry or bad-tempered you are about the fault, remember the service man is there to help you. Help him to help you by welcoming him into the laboratory and by extending the usual courtesies. A chat about the problem in general terms over a cup of coffee allows the engineer to unwind from the journey while assessing the situation. A viable plan of action can then be worked out together with a greater chance of amicable success. This also pays dividends for the future. If your laboratory is one where the engineer feels welcome and at home, his enthusiasm will be greater.* If he has to leave before the problem is resolved but has been made to feel part of your team, his commitment will be enhanced.

Other factors play an important part in motivating the service engineer. It is not pleasant crawling about under a filthy instrument, especially if you find the operator's festering sports kit and a few sandwiches rotting behind the electronics console (oh yes, it has happened to me). Also it is frustrating if there is nowhere to put your tool case or space to work on an instrument subassembly. The job can be brought to a complete standstill if no one is available to help lift a unit out or to hold a meter while the engineer dives inside an awkward module. Just as the service engineer should become part of your team, you and your colleagues should become part of his.

If it transpires that you have given false information about the fault or made a mistake in the readings, be honest about it. Trying to hide your human frailties only leads to confusion about the problem and may get in the way of resolving it. Just as most service personnel try to maintain their credibility ratings with the customers, so too should the customer try to be straight with the manufacturer. Help them to help you.

Do not ask for too many favors. The customers who shout the loudest always seem to get the fastest service. This is true, but they do not always get the best, anything to get him off our backs is a natural response to this type of operator. The good guys, that is those who are firm but fair, get the best results. Remember the instruments are complex. Problems are not always easy to fix. They only become obvious with hindsight. Try to be patient.

The following rules for handling service engineers was pinned to one instrument in a laboratory I visited. They would be funny if only they were not true.

CUSTOMER'S RULES FOR HANDLING SERVICE ENGINEERS

1. Never send for a service engineer until everyone has had time to form their own opinion as to how to fix the instrument.

2. Have a person from each shift (so none will know what the other one has done) work at the instrument and lose as many parts as possible.

3. After sending an urgent request for the service engineer to come immediately, put the gatekeeper wise so that it will be as tough as possible to get into the plant.

4. The minute the service engineer arrives, ask him where he has been and what kept him. Before he can answer, ask him how soon the instrument will be working again.

5. Hide all manuals and schematics connected with the job.

---

*Since male engineers are still in the majority, I use the male pronouns here as a matter of convenience for writer and reader.

6. Always have at least 8 engineers present to ask technical questions which are in no way connected with the trouble.

7. Ask how soon the instrument will be operating again.

8. Be sure to provide a helper who has never seen the instrument before and also be sure that the helper has no tools.

9. The instrument should be as dirty as possible.

10. Tell the service engineer what excellent service other service engineers give their instruments (of course, we realize their instruments almost never require any repairs.)

11. Ask how soon the instrument will be operating.

12. When he finally gets the instrument running, tell him what a swell job he has done although he should have done the job twice as fast.

13. After the service engineer leaves, write or call the manager and imply that the instrument is worse than it was before. This will give the engineer a chance to come again and you can go through the above rules once more.

CHAPTER TWELVE

Fault Finding and Nonroutine Maintenance

There are an infinite number of fault conditions and combinations of failures that could occur in a GC/MS system. In general terms one must follow the troubleshooting procedures outlined earlier. However, certain types of faults and certain situations happen more frequently than others. Some of the more common problems are discussed in this chapter. Armed with this information and a reasonable understanding of the various parts of the instrument, most operators should be able to home in on a problem fairly quickly.

VACUUM SYSTEM PROBLEMS

Gross problems with the vacuum system are uncommon, and when they do occur they are usually easy to define. More common, and often more difficult to pinpoint, are problems such as leaks and contamination. In many cases it is possible to run the instrument in spite of them, and all too often they are ignored. In some cases, the basic lack of understanding of vacuum principles in general and the GC/MS instrument in particular means that faults go undiagnosed. The operator should be aware of good (normal) conditions so that the occurrence of a vacuum problem does not go undetected.

Air Leaks

These are the most common vacuum problem but fortunately the mass spectrometer is a very good leak detector. Air is predominantly nitrogen, followed by oxygen, with carbon dioxide, rare gases, and water vapor making up the rest. With the exception of water vapor, all these gases are relatively easily pumped away by the vacuum system. If there are no leaks into the vacuum enclosure and the system has been pumping for some time, nearly all the residual air will have been pumped away, but water will still be apparent at 17 and 18 u (OH^+ and H_2O^+). If a leak occurs, the levels of water vapor and other gases will rise. As the leak rate increases, the nitrogen ion signals at 14 and 28 u (N^+ and N_2^+) will rise faster than the water vapor ions. The oxygen ions at 16 and 32 u (0^+ and 0_2^+) will rise in proportion to nitrogen in about the ratio 1 to 3, oxygen to nitrogen.

This discriminate pumping of the vacuum system gives us a rule-of-thumb method for identifying leaks and is illustrated in Figure 142. If the ratio of 28 u (N_2^+) to 18 u ($H_2 0^+$) is greater than 2 to 1, then there is definitely a leak and it should be fixed. If the ratio of 28 u to 18 u is less than 1 to 2, that is, if the nitrogen signal is much smaller than the water signal, then the vacuum system is probably leak-free. If the ratio lies between 2 to 1 and 1 to 2, then there may be a small leak or residual air, depending on how long it is since the system was vented to atmosphere. In this circumstance no immediate action is required but the ratio should be monitored. If there is no improvement after a few hours, then it may be worthwhile looking for any small leaks.

There are exceptions to this general rule. For instance, if there is a gross leak, the ion source may saturate at 28 u and distort the displayed ratio. High signals at 14 and 16 u, which are normally much smaller than that of the 18-u ion, and, of course, a high vacuum pressure are clear indicators. A leak a long way from the ion source, especially one that requires the air to travel across a vacuum pump orifice to reach the ion source, tends to give an intermediate ratio. Always monitor the pressure as well as the background spectra, because in this case the pressure would be unusually high.

Finding the source of the leak should be straightforward, but many people seem to have difficulties. Leaks rarely occur for no reason. Always consider what has just happened to the instrument. Has the source just been reinstalled or a new column fitted? Has the chromatograph oven or interface cooled, causing a ferrule to open up? Make use of diverters and the variable-geometry source, if fitted. In other words, reroute the gas flows or change the pumping speed in parts of the instrument to try to locate the basic leak area.

Probe for the leak with a chemical tracer. A cotton swab dipped in methanol is usually quite good. Tune the mass spectrometer to 31 u and look for a rise in signal level. Be ready to switch off the filament and multiplier if the leak is fairly large, as a strong solvent signal will be evident. Often it is a good idea to reduce the filament current and multiplier voltage anyway when dealing with large leaks. When checking for leaks this way, one often sees a drop in the air signals when the methanol initially blocks the hole. With very large leaks, monitor the ion gauge response rather than use the mass spectrometer.

Use methanol in preference to other solvents as it is less likely to damage plastic and paints. Nevertheless, use it carefully. Never squirt solvent on the instrument, for it might cause a serious fire. Better than methanol is a small gas

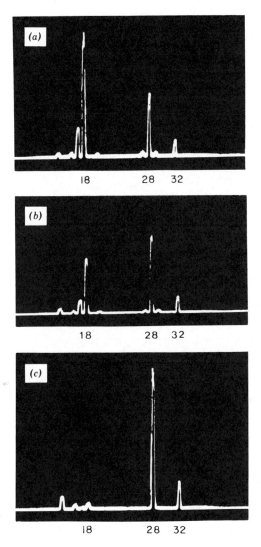

Figure 142. Air related spectra. (*a*) No leak, residual air and water. (*b*) Small air leak. (*c*) Gross air leak. Note that the vertical scales are not the same. (Copyright © 1982, Finnigan Corporation. All rights reserved)

bottle of argon with a tube terminating in a fine jet. This can be extremely accurate in pinpointing leaks.

Monitor the argon molecular ion at 40 u (Ar^+) before opening the test bottle to check the background plus-leak signal level. If the leak cannot be found but the 40 u ion signal slowly increases over a period of 5 to 10 min, then there must be a leak from the laboratory that is now enriched with argon from the test bottle. Backing line leaks and backstreaming rotary pumps (low oil levels) are typical causes of this sort of response.

It is extremely difficult to find leaks on backing lines; solvent and argon are not easily seen unless they are used in large quantities. You may be able to detect methanol if you soak a cloth or tissue with solvent and wrap it around areas of backing line, but this technique is potentially dangerous and is not recommended. It is better to use a Freon-type spray if it is permitted. The types of spray sold for cleaning disk and tape heads (essentially trichlorotrifluoroethane), rather than the freezing sprays (fluorinated methane), are very effective in backstreaming through diffusion pumps from a backing line leak. Look for chlorine isotope patterns at 85, 87 u and 120, 122, 124 u. The freezer-type Freon sprays are even more effective and they pump away more rapidly, but their rapid cooling action can be the cause of additional leaks at backing line connections.

Virtual Leaks

If the air leak cannot be located by these methods, then it may be a virtual leak, a pocket of air trapped inside the vacuum enclosure and slowly leaking into the main pumping area. Virtual leaks will eventually pump away, since the air reservoir is finite. Any system that has been pumping for more than a day is unlikely to have a virtual leak, unless there are some very large connected chambers or tightly fitting enclosures in which large air pockets can be trapped. The inside of a CI reagent gas selector with poorly sealing selector valves is a typical example. The space around a solids probe inlet between the inner and outer valves is another. These can usually be pumped out by an auxiliary pump and soon eliminated from the search.

Gas-line inputs, the helium carrier gas line especially, can be a major cause of virtual leaks, the air being mixed with the gas in the line. This may be due to long lines between the GC/MS and the gas cylinder, which take a long time to purge if either has been opened, or there may be a leak in the line itself. There is a mistaken belief that because gas lines are at pressure air cannot enter. This is, of course, not true. Air can easily diffuse into the line, causing a large partial pressure of air in the vacuum system. If blanking off a gas line, disconnecting the column for instance, reduces the air spectrum, then you must leak check with a solution of soapy water along the gas lines. If gas is getting out, then air is getting in.

Pressure Limits

Sometimes you are faced with poor pressures but background spectra typical of a leak-free system. Detailed examination of the background usually reveals a higher total signal level than normal. The pumps are probably working inefficiently, giving a poor ultimate pressure. In the case of rotary pumps, the oil level may be low, the gas ballast valve may be open, or one of the backing pumps may be out of action. This is also true for diffusion pumps, which can also show reduced pumping speed if the oil is contaminated or the pump is running too cold. This can be caused by a faulty heater, a low main power supply voltage, or a very cold cooling water supply. In the case of turbomolecular pumps, a drop in the speed can reduce the ultimate pressure obtainable.

Diffusion pumps can also show some interesting phenomena if their oil is dirty, the level too low, or the heater inefficient. The pump oil may boil erratically, causing fluctuations in pumping speed with a corresponding change in vacuum pressure. The effect is not always sudden. It is not uncommon for pumping systems to cycle up and down in pressure over 5 or 10 min as regularly as clockwork because an incorrectly rated diffusion pump heater was fitted or the line voltage is outside the normal pump specification. A similar effect can occur if the heater has a few shorted turns that make intermittent contact with each other.

Sometimes manufacturers of GC/MS systems use a particular pumping fluid for which the diffusion pumps were not specifically designed. The new fluid may have different boiling characteristics than the old, and the correct fluid charge may be different from that specified in the pump handbook. Other factors may influence the amount of fluid actually required. For instance, pumps designed for use with silicone oils are sometimes filled with polyphenol ether oils, silicones being undesirable in the main GC/MS pumps. To achieve satisfactory boiling characteristics the pumps may be underfilled by up to 20%. Alternatively, there may be a tendency to lose oil onto cold baffles and the underside of isolation valves, and a 10% overfill is required. Always consult with one of the GC/MS manufacturer's service or production engineers before refilling diffusion pumps; the required levels may not be as stated in the pump handbook.

These are fairly rare faults. High pressures are often caused by quite obvious partial pressures of known gases, such as helium or CI reagents. One expects these gases to be in the spectrum at a high level, and there is a tendency to discount their presence. Always check the flow rate settings, or consider a leaking ion source in the case of CI if the pressure is abnormally high. Contaminants can also raise the pressure, but they are not expected and their spectra are not so easily overlooked.

Contamination

Once again the mass spectrometer is both doctor and patient. Contamination from whatever source will have a characteristic spectrum that can be identified, and once identified the cause of the contamination can usually be located. The most common contaminants are the pumping fluids of the system itself. Figures 143 to 145 show typical spectra for mechanical pump oil, polyphenol ether diffusion pump oil, and a silicone diffusion pump oil. The ions in the latter are also typical of vacuum greases often used (excessively) to lubricate Viton 0-rings before vacuum chamber assembly or to ease moving seals such as those on solids probe inlets.

A strong signal at 149 u is indicative of phthalates, industrial plasticizers that

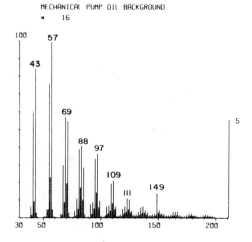

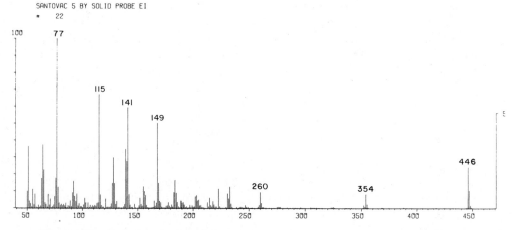

are found in most plastic materials in a variety of forms. Figure 146 shows the spectrum of dibutyl phthalate. It is a poor mass spectrometer (in a GC/MS system) that does not show some signal at 149 u, such is the ubiquitous nature of phthalates, and the signal is a useful background marker. However, a strong signal may indicate contamination from an overheating plastic tube, a new septum, or even the residue from plastic or synthetic rubber gloves following a source cleaning. The level should reduce quickly, especially if the vacuum chamber is warm. If it persists, look for an overheating backing line tube or a piece of glove or plastic bag inside the vacuum system.

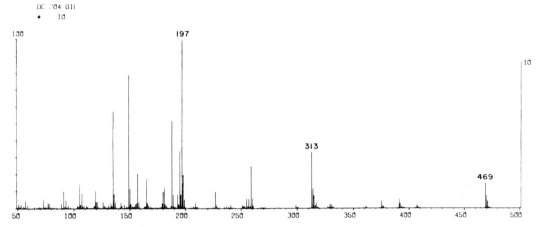

Figure 145. Silicone diffusion pump oil, EI spectrum. (Copyright © 1982, Finnigan Corporation. All rights reserved)

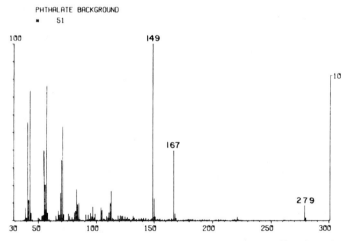

Figure 146. EI spectrum of phthalate background. (Copyright © 1982, Finnigan Corporation. All rights reserved.)

Occasionally, one sees a series of ions 16 u apart, corresponding to the oxides of rhenium. The series starts at 185 and 187 u, the two isotopes of rhenium, and extends to 233 and 235 u (ReO_3^+). In addition, a stronger ion pair at 250 and 252 u corresponding to $HReO_4^+$ is also seen (Fig. 147). In fact, with CI systems the perrhenic acid ions are sometimes seen even when the oxide ions are not particularly obvious and may be due to a CI plasma containing a significant amount of water and/or oxygen attacking the filament.

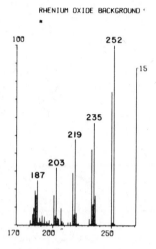

RHENIUM OXIDE BACKGROUND

Figure 147. EI spectrum of rhenium oxides. (Copyright © 1982, Finnigan Corporation. All rights reserved.)

Damaged GLT Lines

Another interesting ion signal to watch for is 446 u. Not only is this found in diffusion pump oil spectra but it is also the main ion from the adhesive used to bond glass to metal in glass-lined tubing. If the GLT has been cracked, the glue will seep into the mass spectrometer and a stronger 446 u ion may be seen. Even so, the ion intensity is usually very low but becomes more obvious whenever the GLT lines are warmed up. Always look for the 446 u ion when changing interface temperatures to higher settings.

Apart from these common contaminants, thousands of other compounds may be seen. The usual causes are dirty gas lines or a bad gas tank. Even clean and highly rated cylinders are apt to give a dirty supply as their pressure drops near the end of their useful life. Contamination is often introduced by the system operator either by accident or by bad practice. This may be at the sample preparation stage, at an inlet to the system (dirty syringes and dirty probes), or even within the vacuum chamber. I once stripped a vacuum system that was giving some problems, only to find a cigarette butt inside the source manifold! There are many reasons why one should not smoke in a GC/MS laboratory, but I had never thought that a poor background spectrum would be one. More obviously, there is a tendency to get careless or perhaps carefree, when handling the "works." A fingerprint can give quite a confusing background spectrum. Make sure the vacuum chamber is covered when parts are removed for cleaning. It is all too easy for a fly to get into the system.

A lot of contamination, of course, is quite deliberate; the muck is put there as part of a sample under analysis. The cleaner and smaller the sample, the better the state of the vacuum system. Running the manifolds warm, and in some cases a bakeout, helps to reduce the effect of contamination due to samples. Contamination from the gas chromatograph either as column or septum bleed is also inevitable. Running these as cool as possible, cooling the column oven when it is not in use, and using low-bleed materials and lightly loaded columns all help to keep a low background. Get to know the bleed ions for a given column so that new ions coming from the gas chromatograph inlet can be quickly spotted even if not immediately identified. In short, examine all the background, not just the mass range of immediate interest; the mass spectrometer may be trying to tell you something.

CHROMATOGRAPHY AND INTERFACE PROBLEMS

The quality of chromatography can vary enormously from one laboratory to another. What is acceptable in one place may be totally inadequate elsewhere. Always strive to achieve better chromatography, as this will improve the overall system performance in terms of sensitivity and detection limits. It will also increase the likelihood of separating two or more co-eluting compounds and thereby increase the chances of valid identification. In other words, do not just accept your present standards, but try to improve them. Always be aware of any reduction in standards or quality, as these are a sure sign of problems.

Chromatographic Response

The detected peaks should be smooth and symmetrical and as narrow as possible. Any tendency for the peaks to broaden, tail, or split is indicative of chromatography or interface problems. Peak broadening, loss of resolution, can be caused by column packing, that is uneven, loose, or discontinuous packing. Any sudden increases in the diameters of fittings and interface lines can have a similar effect.

Tailing is a nonsymmetrical chromatographic peak with a continuing response for a longer time after the peak top than before. It can be caused by the same things as peak broadening. Void volumes or dead volumes that are sometimes found around ferrules and fittings and in injectors can also give this effect. Cold spots, which cause sample condensation, also give this type of response, as do active sites; which are areas within the column or transfer lines that can absorb, delay, or degrade the sample. Active sites are often caused by contamination within the chromatographic system, such as pyrolyzed samples, by exposed support material where the liquid phase has been stripped off, or by exposed metal surfaces, such as broken GLT lines or in ferrules and fittings.

Splitting of the chromatographic peaks may be the result of uneven gas flows in the column and can be caused by leaking connections. Alternatively, cold spots, especially at the injector end of the column or the injector itself, can give the equivalent to multiple injections. A poor injection technique could have the same effect. Figure 148 illustrates some of these problems. It was noticed that the solvent peak of a particular GC/MS system was very broad. Closer examina-

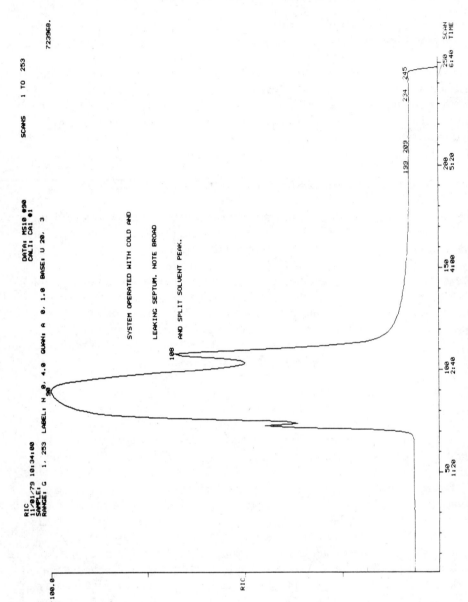

Figure 148. Effect of a cold injector and leaking septum on chromatographic response.

tion showed that the solvent peak was also split. Investigation of the gas chromatograph revealed a cold injector and leaking septum. The system was being used for very low level detection work, and no chromatography peaks were seen. It would have been easy to assume that the sample was a blank if no one had noticed the unusual solvent response.

Another chromatographic problem, especially common with capillary work, is condensation of contaminating material onto the column. As the column is heated, this material elutes as poorly resolved peaks that interfere with the sample response. This is illustrated in Figure 149. A sample of plant extract was analyzed on a capillary GC/MS system. The initial run (upper RIC trace) was made first thing in the morning. The column had been left connected overnight with gas flowing through it at room temperature. Significant contamination from the gas lines, very long copper tubes in this case, had condensed on the column. At the end of the run, the column was held at its upper temperature for about 20 min and cooled quickly, and the sample was rerun. As can be seen from the lower trace, performance had noticeably improved. Fitting filters at the GC/MS end of the gas lines can help reduce this type of problem.

Fittings and Ferrules

It is very important in chromatographic work to choose the right fittings and ferrules for the job, or dead volume problems may occur as discussed earlier. To prevent sample degradation it is also important that as little of the fitting's metal surfaces be exposed to the gas stream as possible. Fittings, ferrules in particular, have upper temperature limits that should not be exceeded. Even when ferrules are operated within their temperature limits, leaks can occur, especially in areas where the temperature is cycled up and down. Leaks result in sample loss and can reduce sensitivity as well as causing chromatographic problems. They also allow air to diffuse into the system, which can degrade the column packing and cause active sites. Oxygen in the mass spectrometer is also undesirable.

Ferrules have a tendency to stick to the pipes and in some cases actually deform the pipe surface. Do not overtighten fittings, or damage may occur. With delicate lines, such as GLT, metal ferrules should not be used. Obviously, with glass tubes such as columns, metal ferrules cannot be used, but they should not always be the first choice for some parts of the metal pipework either. A lot of GC/MS assemblies, such as source flanges, are complex structures that include short lengths of gas inlet and pumpout tubes welded into them. If the short length of pipe is damaged, the whole assembly may have to be scrapped. The use of a softer ferrule in these areas is recommended.

Syringes

A good syringe technique is essential for satisfactory GC/MS work. Syringes are delicate pieces of equipment and should be looked after carefully and checked frequently. Blowback past the needle can be a cause of sample loss. To check for

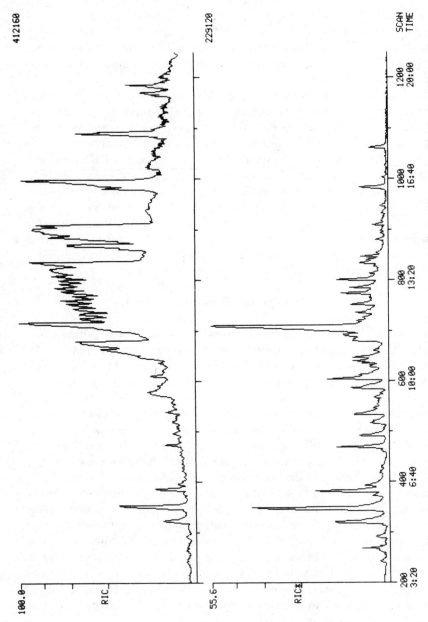

Figure 149. Contamination condensed in the column spoils the chromatographic response (upper trace). Rerunning the sample after holding the column at high temperature for about 20 min gives a greatly improved response (lower trace).

this, inject a small amount of solvent into an operational gas chromatograph, but delay pressing the plunger after the needle is inserted through the septum. Hold a finger over the plunger to stop it from blowing out. If there are any leaks back along the stem of the syringe the solvent will be pushed out of the barrel. A few air bubbles in the syringe help to detect this fault.

Bent and blocked needles are also a common problem. Some syringes have replaceable needles. For syringes that do not, the needle may be straightened out and the syringe then tested for satisfactory performance. Blockages can sometimes be forced out by pushing down the plunger if the syringe contains enough solvent. On some designs it is possible to wet the plunger stem with solvent and then withdraw it from the body. The reduced pressure inside causes the solvent to rush into the barrel. The plunger can then be reinserted and the blockage pushed out. Do not use excessive force, or the syringe barrel may crack. Always check the syringe for blowback after clearing it in this way.

Syringes can be a major source of contamination from both inside and outside the needle. Always rinse the syringe thoroughly between samples, and be extra careful to clean it completely when it is put away and before it is used again. Samples can be concentrated by the solvent effect within the syringe needle, and it is advisable to run a confirmatory blank solvent sample when doing low-level work or when moving from concentrated to less concentrated samples.

Each operator has his or her own injection method and favorite syringe. Providing the techniques adopted are consistent, the chromatography should be repeatable and reliable, and any deviations from normal should be immediately detected. Do not dismiss a poor chromatography result as a "bad injection." Develop a consistent technique so that the injection is not a variable in your chromatographic work. Always investigate poor results.

Blockages

It is uncommon for a packed column to become blocked, but a new column may be too tightly packed and gas flow may be restricted. A high head pressure at the injector for a given flow rate is an indicator of blockage. Normal parameters for the system should be established so that faults will be easily detected. Apparent blockages can be caused by pushing the column too far into the injector so that it seals against the underside of the septum. Pulling the column back about a millimeter usually clears the fault. Grinding a groove across the top of the column is a good way to prevent this and allows you to get the column in the same position so it can then be pushed right up to the septum. Some injector designs have an internal step to prevent the column end from touching the septum. However, none of these methods is foolproof, as the septum may extrude downwards in use and fill the end of the column, creating a restriction.

Interface lines, especially jet separators, are far more likely to block. Complete blockages are apparent from the changes in column head and source pressures. Partial blockages are not so easy to detect. There may be some pressure changes, and the diverter response—the response of the mass spectrometer when the diverter is used—may be sluggish. Partial blockages of the separator negate its enriching action, and sensitivity may be greatly reduced. The separator should always be checked if loss of sensitivity is a problem.

It often happens that a separator will block again soon after cleaning. Examine the separator under a microscope after cleaning and check that there is no particulate matter such as column packing in the jet stems. This will be blown back into the jets as soon as the system is pumped down. Also check that the filter frits fitted at either side of the separator arms are in good condition and that the fittings are clear. Small pieces of ground glass or broken frit caused by overtightening the fitting can be the cause of the problem.

Also check interface temperatures when blockages occur. Cold spots or a faulty heater control circuit may cause sample condensation. If blockages occur frequently, check the temperature of the interface independently as the display may be giving a false readout. The interface or part of it may be running at a lower temperature than indicated.

Sensitivity

If the mass spectrometer response seems lively and sensitive to background ions (and for solids probe samples) but GC/MS sensitivity is a problem, then the column must be suspect. Always check the system with known standards that have been properly stored. A standard that is concentrated due to evaporation of solvent will give a false expectation for the instrument. On the other hand, the standard may have degraded.

For instance, on a number of occasions I have found that a methyl stearate solution was not as strong as it was when freshly made. On the other hand, the palmitate impurity appeared just as strong or even stronger. This is observed as a peak eluting just before the stearate with strong 74 and 284-u ions, its base peak and molecular ion. These changes seem most pronounced when the methyl stearate is made up in acetone. I do not feel qualified to say that the stearate degrades to palmitate. Certainly a number of chemists have told me that this will not happen. Nevertheless a change has been noted by myself and others in standard solutions. The only safe conclusion is that a fresh solution should be made if there is any doubt.

In other cases the standard sample may be absorbed by the sample bottle's plastic cap seal. Once again, if there is any doubt, make up a fresh solution. If it is confirmed that the sensitivity is low, check the column and interface system with regard to the points discussed earlier. Also check the system with less labile compounds such as a straight-chain hydrocarbon. Differences in expected response may indicate active sites within the system.

For instance, exposed metal in a fitting or transfer line can drastically reduce sensitivity for some compounds. Figure 150 shows the effect of a stainless steel transfer line compared with glass-lined tubing. The upper trace shows the 74-u mass chromatogram, and below it is the RIC response for a 10-ng injection of methyl stearate. A small peak at scan 50 indicates the presence of methyl palmi-

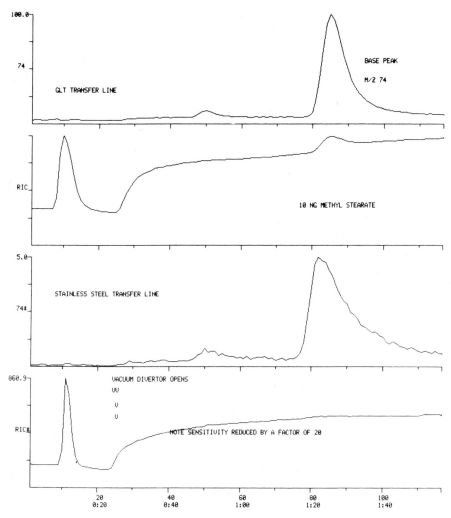

Figure 150. Comparison of stainless steel and glass-lined tubing used as a GC/MS interface transfer line.

tate. The lower traces show the same data obtained with a stainless steel transfer line instead of a glass-lined tube. The mass chromatograms are drawn to give full-scale deflection, but the scale indicates only 5% response for the 74-u ion when the stainless steel transfer line was used, a 20-fold reduction in sensitivity. The peak cannot be seen at this level on the RIC trace.

It is also interesting to note that although a vacuum diverter was used in both cases, a solvent peak can be clearly seen, showing that the diverter is not 100% efficient. The relatively fast recovery of the diverter implies that the separator is functioning normally and is not the cause of the loss of sensitivity.

The effect of active sites in a standard GC/MS system is clearly shown in the analysis of cholesterol. If the interface temperatures are too hot or there are any catalytic surfaces, then the cholesterol tends to dehydrate to cholestadiene, and the two compounds co-elute. They may separate on high-performance systems if the degradation occurs at the injector end of the column, but it is more common to have a peak with tailing. Examination of the spectrum shows a relative increase in the $M-18$ ion at 368 u. In severe cases the 368-u ion may even be the base peak.

Systems that are relatively inert will give a good chromatographic peak with cholesterol, and the molecular ion at 386 u will be at least twice the signal at 368 u as shown in Figure 151. Some manufacturers quote even lower figures, but these may be a result of ion source settings as well as due to the column-interface condition. Cholesterol is a good sample for checking the GC/MS system and is used by many laboratories. If possible, run a sample by solids probe first. This will give the 386 to 368 u ratio for the source conditions alone. Take care not to saturate the mass spectrometer by using too large a probe sample, or a distorted ratio will be recorded. Any reduction in the ratio of the two peaks when the

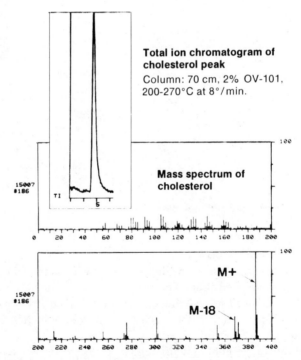

Figure 151. Chromatogram and spectrum of cholesterol. There is no tailing, and the molecular ion at 386 u is much greater than the $M-18$ ion, indicating a chromatographic system free of active sites. (Hewlett-Packard Corporation)

sample is passed through the gas chromatography inlet indicates problems with active sites.

Saturation effects are a common problem in GC/MS analysis of complex mixtures. At best, they stem from the wide range of component concentrations in the sample; at worst, from careless or haphazard injection techniques. Operator mistakes are a major cause of chromatography problems in GC/MS work and can waste a lot of instrument time. This is illustrated by the traces in Figure 152 taken during the analysis of a plant extract by capillary GC/MS. The upper trace gave satisfactory spectra for most of the peaks, but some are clearly overloading the column and saturating the detection circuits. In our haste to rerun the sample at a lower concentration, we forgot to reset the automatic solenoid valves that controlled the Grob injector. The second trace was therefore useless before scan 800, and the sample had to be run a third time, wasting about an hour in all. One does not always have a lot of sample or time, so care and attention to detail is most important.

MASS SPECTROMETER PROBLEMS

Sensitivity

Apart from major hardware failures, the most commonly reported problem with the mass spectrometer section of a GC/MS system is poor sensitivity. It is often unnecessary for the operator to run samples before deciding that the mass spectrometer is insensitive, he gets a feel for the current status by looking at the background spectrum, the noise level, and the GC/MS data. In fact, it is very hard to devise a simple test to check the mass spectrometer sensitivity independently of the rest of the system, and one should take care not to jump to the wrong conclusions. It is possible that the background levels have changed. Check the vacuum pressures. If the loss of sensitivity is fairly sudden, check for the obvious, silly things that are so easily overlooked. Has someone changed any of the source parameters and altered the tuning? Has the multiplier power supply been turned down? Has a filter been inadvertently set to a limiting value? It is worth a few minutes to check the obvious before stripping the mass spectrometer down or calling for assistance.

If the loss of sensitivity has been gradual, then it is likely that the source has been slowly contaminated, a normal process in GC/MS operation. Gradual changes in the tuning parameters will support this theory. Alternatively, the multiplier gain may have fallen off, and a cleaning, rejuvenation, or replacement may be necessary. If possible, check the multiplier gain, but take care not to introduce troubles by subjecting it to an excessive ion current. In most cases, ion source contamination and multiplier aging go hand in hand. The multiplier gain can be restored by increasing the power supply voltage, but an increase in noise level must be expected. The source must be cleaned to restore its performance. After cleaning the source or assembling any other part of the mass spectrometer,

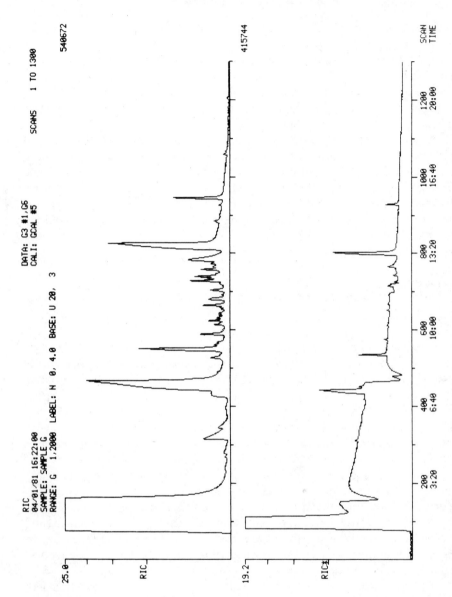

Figure 152. More haste, less speed! The top trace has overloaded peaks. In the rush to rerun the sample (lower trace), the split and sweep solenoids were not reset.

you may find that the sensitivity is not restored and it may even be worse. Alignment or the lack of it may be the problem. Pay particular attention to this when reassembling the mass spectrometer components.

Charging Effects

Difficulties in tuning the mass spectrometer and an unstable spectrum are also common problems. They too can be caused by contamination. Any tendency for the spectrum to float up and down—that is, a slow response to changes in the tuning parameters—indicates charging. Contamination on part of the mass spectrometer may form an insulating surface on which charge can build up. This deflects the ion beam, defocusing the instrument. If the charged surface is a lens plate, the effect of altering the lens voltage will be dominated by the trapped charge, giving a sluggish, spongy response.

Parts of the flight tube, on either a sector or quadrupole instrument, may also charge up and reduce ion transmission. This type of problem is more difficult to detect. If the supply of charge is the ion beam itself, then the effect is made worse when operating with high ion signals, such as ions from an air leak or a CI re-agent gas. If the charge is due to leakage from one of the high voltage supplies, there may be other indications such as a limited or changed mass range. Always check insulation of the mass spectrometer components carefully during reassembly, after cleaning, or when problems are suspected. Most test meters apply only a few volts to the terminals, and breakdowns cannot easily be detected. On the other hand, leakage resistances of a few megohms can usually be seen. While this may not seem like a short circuit, at 5 or even 10 kv volts quite a current will flow and the leakage path may even break down to a lower resistance. Always check for short circuits with the meter set to detect high ohm values. Take care not to measure the resistance of your fingers; use insulated probes. In this way true short circuits will be found and leakage resistance will not be overlooked.

Another charging effect can occur if the ion current is high. The concentration of ions at lenses and slits can cause space-charge effects that can defocus the ion source. Such situations can occur when operating in the CI mode or if there is a large leak in the system. In the former case, improvements may be possible if the reagent gas pressure is reduced. Although there are accepted pressures for optimum CI operation, the displayed pressure may not truly reflect conditions within the source, and the operator should experiment with settings if difficulties are encountered.

Space-charge effects can also occur if there is a strong ion current signal, possibly due to an air leak. Never operate with large air leaks even though you are looking at ions well away from the air peaks. Bunching of ions at the source slit or a strong ion current to the quadrupole can significantly affect tuning and sensitivity. A similar effect can occur at the collector slit on sector instruments or quadrupole exit aperture when you are monitoring a very strong ion. Once again, the advice is to use as little sample as is necessary.

Pressure Effects

Another problem to watch out for is charge exchange spectra, due to the high pressure of helium when it is used as a carrier in direct connections to the ion source. In the EI mode, you may be able to increase the source conductance by opening slits or apertures. This is not possible in CI, and you should add the reagent gas as makeup to ensure the best conditions possible.

Poor CI tuning and sensitivity may be caused by the loss of the CI reagent gas due to leaks in the ion source. This is difficult to detect, but one sometimes sees slight changes in source and analyzer pressure relationships. Fluctuations in the pressures may also indicate gas leaks from the source. Variations in the pumping speed in EI or CI can also be the cause of unstable spectra, and the pumping system should be checked.

Filament Problems

One of the most obvious instrument faults is filament burnout. If there is no ion signal, check the filament first. Check for continuity between the filament connections. Set your test meter to low ohms, correctly zeroed, for this test. When cold the filament will be less than 1 Ω on most instruments. A low ohmmeter setting is needed to detect if lead and connection resistance has increased. Even a few ohms in series with the filament may prevent it from working. The driving power supply may have only a few volts available even though it usually gives a relatively high current, 4 A typically, to heat the filament.

Many instruments have a view port so that the filament glow can be seen. If there is filament continuity but no glow, check that the supply leads are not shorted together. Also check for shorts to the manifold and other parts of the ion source. Set the test meter for high ohms for this test so that leakage paths are easily spotted.

Erratic filament operation can be caused by a number of faults. One of the power supplies may be unstable or incorrectly set. Check the electron energy voltage and the ion energy voltage as well as the filament drive power supply. There is usually a complex interaction between the supplies to control and regulate filament emission. Abnormal pressures and misalignment in the ion source can also cause erratic filament operation. A bent filament will not work in some instruments, because the electron entrance holes are very small. Alignment is critical. Always check this when assembling the ion source, especially when using repaired filaments. Check also that the filament wire or ribbon is taut; it may sag when hot. Sagging can cause alignment faults and intermittent short circuits.

Small magnets are often included in ion sources; larger ones may be mounted externally. These are used to spiral the electron beam to increase the probability of an electron-molecule interaction. Loose or incorrectly positioned magnets are an additional cause of filament problems. Often one adjusts the magnet position to optimize the electron transmission across the source. A high trap current (in sector instruments; called collector current on some quadrupoles) relative to the total emission is indicative of a clean source. However, the ratios of emission to collected current are not as important as mass spectral patterns. The position of the ion source magnet can have quite an effect on overall ion source tuning and the resultant mass spectral high mass to low mass ratios.

As the source gets dirty, filament current emission can be difficult to regulate if trap (collector) current control is used. In most cases, and always in CI, emission current regulation is preferred. It may be necessary to increase the emission level on a dirty source to compensate for poor electron penetration in the ion volume. On instruments set for trap (collector) current regulation, a dirty source may be the cause of filament instability problems.

Resolution and Mass Peak Shape

Like sensitivity, peak shape can be greatly affected by contamination and charging effects. In trying to improve peak shape, resolution as well as sensitivity may be lost. Peak shape on sector instruments is primarily a function of slit sizes and will be affected by any dirt on the slit plates. A small hair or piece of fluff sucked into the vacuum system when it is vented can easily get stuck on a slit assembly and give rise to jagged, split, or even double peaks.

Resolution problems often indicate an unstable or noisy power supply. On sector instruments, either the accelerating voltage and/or the electrostatic analyzer voltages may be unstable or the magnet control circuits may have faults. With quadrupole instruments a fault with the rod dc voltages is a possibility. On the other hand, the problem may be one of adjustment rather than a hardware fault. The fringe fields at the entrance and exit of the magnet on sector instruments can cause the image to curve, reducing effective resolution. The image-curvature correction or Z-length correctors are generally set up for the instrument and then left alone. Similarly, there may be a facility to rotate the collector slit slightly to correct for magnet fringing effects. These settings may have been altered or vibrated to a new position.

Mass spectrometers are very susceptible to stray fields. Check whether there are any magnetic materials or new pieces of electrical equipment near the mass spectrometer. Where there are two sector-type mass spectrometers in the same room, it is normal to position their magnets at right angles to minimize cross talk. Even so, it is still possible to detect the presence of a second instrument 5 to 10 m away, especially when operating at high resolution.

All types of mass spectrometers will show loss of resolution at high scan speeds. There is a direct tradeoff between scan speed and resolution with sector instruments. Quadrupole instruments can normally be operated at much higher speeds than sector instruments and are usually set for unit mass resolution (separation of adjacent nominal mass numbers) so that resolution–scan speed problems are not common. It is possible to exceed the instrument limits, though, especially in the MID mode. More common, however, are peak-shape limitations and effects caused by amplifiers and filters. Always check whether the cor-rect filter and settling times have been selected.

Quadrupole instruments are critically dependent on the performance of their rod structures. Any defects in the rods can have a very marked effect on peak shape and resolution. As the rods become contaminated, the base of the mass peaks tends to broaden, making baseline resolution difficult. This is usually as-sociated with a higher than normal ion energy voltage ("quad. offset" on some instruments). A very high ion energy voltage with poor peak shape and sensitiv-ity usually indicates severe contamination or charging effects on the rods. Also common with quadrupole instruments are ragged peaks with splits or precursors (Fig. 153). These are nearly always caused by localized dirt or dust particles, which are often left after cleaning. The rods must be scrupulously clean to be effective.

All quadrupole rods are an approximation to give the required theoretical field pattern. Dirt, the wear due to cleaning, and ion burns all cause defects that reduce ion transmission and affect sensitivity. It is worth reversing the applied voltage or even turning the quadrupole around in its holder. It is not uncommon to gain a tenfold improvement in sensitivity just by interchanging the two rod voltage leads at the analyzer head. Always mark the leads and note the position of the quadrupole in its mounting and the setting of the polarity switch if there is one. A change in one of these may be the cause of a sudden change in sensitivity or peak shape.

Mass Discrimination

The term "mass discrimination" is most commonly associated with quadrupole instruments because they have a fixed mass peak width instead of fixed resolu-tion, and thus the term refers to a feature of the instrument type rather than a fault. However, discrimination faults do occur, especially if the ion energy volt-age is already set so high that programming up may not be possible. The prob-lem may be one of crossed controls, with one parameter set too high or too low and another set out of normal limits to compensate. Always go back to the last known set of good instrument parameters if tuning is a problem. Some operators adjust the controls, looking only at mass peak amplitude and not peak shape. It is also necessary to consider the relative effects of each tuning adjustment at both high and low mass. This can be done on some systems by displaying a high-mass and a low-mass ion simultaneously while the instrument is being adjusted.

Mass discrimination on quadrupole systems may also be due to the x-ray shielding between the multiplier and the quadrupole exit aperture. The instru-ment relies on the multiplier voltage generating a sufficient field pattern so that the ions leaving the quadrupole space are caught and drawn around the optical barrier of the x-ray shield. If the voltage is too low, this will be inefficient. A voltage that is sufficient for ions of 200 u may be several hundred volts too low at 500 u. For operation at higher than normal masses, it may be necessary to in-crease the multiplier voltage to get effective ion transmission. All this comes to

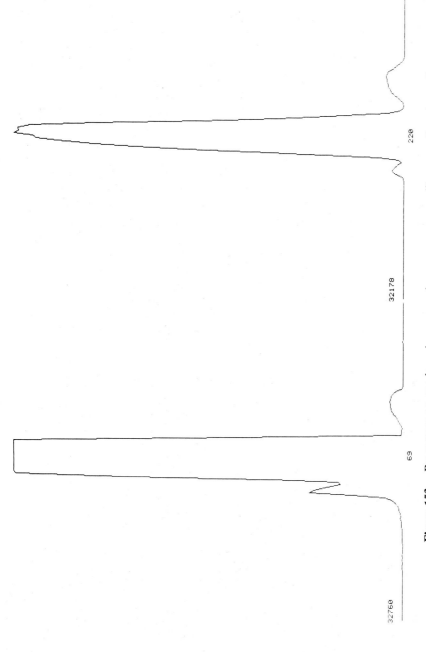

Figure 153. Precursors on quadrupole mass peaks can appear as splits or small peaks on the leading edge of the main peak. The small peaks after the major peaks are due to the ^{13}C isotope and are normal.

319

naught if the x-ray shield is contaminated, as illustrated in Figure 154, for it can then charge up and repel the ion beam.

Figure 155 illustrates the effect of cleaning a dirty x-ray shield and increasing the multiplier voltage. The operator had complained that sensitivity of his Finnigan 4000 was very poor, at higher masses. The laboratory used cholesterol as a standard. The problem was not due to chromatographic effects degrading the cholesterol, as the 386 u to 368 u ratio was better than 2 to 1. Increasing the multiplier voltage by 200 V gave about 10% improvement in the ratio of the molecular ion at 386 u to the low-mass peaks. Cleaning the x-ray shield, however, made a vast difference. Figure 156 shows the improvements with one of their samples. Detail that was previously lost in the noise can be clearly seen in the 400 to 500 u range.

Similar charging effects are also possible on sector instruments when the collector slit is contaminated, but this is not common. Mass dependence on source tuning is the usual cause of mass discrimination faults in sector instruments. The faults become more pronounced at lower accelerating voltages and can occur in voltage scanning. For wide mass ranges, magnet scans are preferred. If voltage scanning is necessary, such as in MID work, try to keep the accelerating voltage as high as possible. For instance, it may be just possible to get a 2 to 1 mass range by scanning from 5 to 2.5 kV at a given magnetic current. Scanning from 4 to 2 kV at a slightly lower magnetic current may not give the same performance at the high end of the mass scale due to the reducing efficiency of ion extraction and focusing at the lower accelerating voltages.

DATA SYSTEM PROBLEMS

A data system is a very complex interdependent network of circuits and mechanical assemblies. When faults occur, there is often very little the operator can do

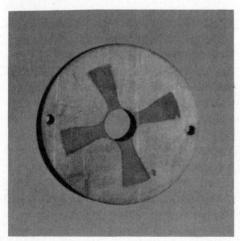

Figure 154. Contamination on a quadrupole instrument's x-ray shield assembly. The material was almost transparent under normal lighting and only became clear when the quadrupole exit aperture plate was dipped briefly in acid to highlight the cross shaped pattern caused by the build up of dirt.

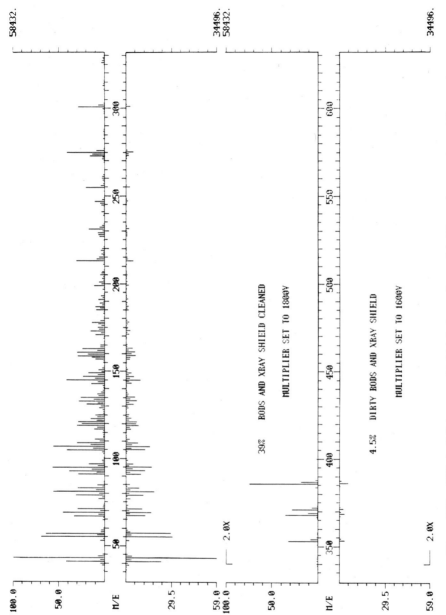

Figure 155. Effect of dirty x-ray shield (cholesterol sample). The rods were cleaned first and the multiplier voltage raised but cleaning the x-ray shield had greatest effect in this case. Improved high mass performance was achieved by cleaning. All three should be checked when ratios are poor. Note good chromatographic performance as the 386 u ion is more than twice the 368 u ion. Lower trace, before cleaning. Upper trace, after cleaning.

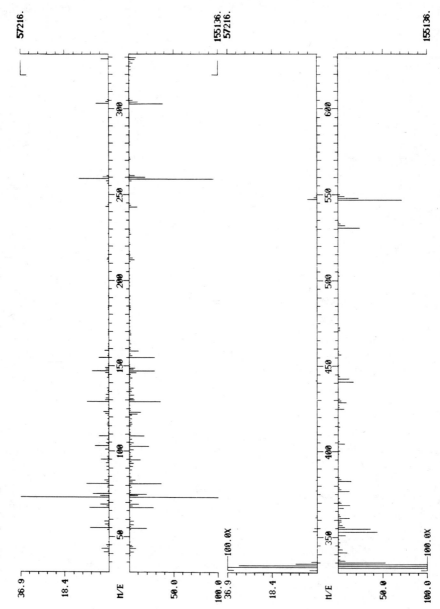

Figure 156. Effect of dirty x-ray shield on a typical sample. Upper trace, before cleaning. Lower trace, after cleaning. Note progressive improvement at high masses.

322

other than change the defective module. The problem is to locate the defective unit. Sometimes this appears clear-cut—a disk unit will not become ready or a tape will not rewind, for instance—but one should not jump to conclusions. Many data system units have elaborate hardware (physical and electrical) interlocks as well as software (program) constraints. It may be that these are preventing normal operation. Where things appear normal but a fault is suspected, the operator must use software diagnostic test procedures to exercise each part of the system in turn to locate the problem area.

Diagnostic Procedures

Nearly every part of the data system will have one or more diagnostic test programs associated with it. These will have been developed by the computer or peripheral manufacturer and are usually provided by the GC/MS manufacturer in the standard system software or as separate test software. The procedures range in complexity from three-line programs entered at the computer console to elaborate quality control exercisers. In all cases, the operating procedure must be followed exactly or spurious false starts and errors will be generated. The test procedures are generally detailed in the data system service and maintenance handbooks that are available from the GC/MS manufacturer. While the procedures are straightforward, they can be confusing to the computer layperson, and it is recommended that the GC/MS operator either attend a customer training course or arrange for some basic on-site training before faults occur.

Some of the diagnostic routines are quite comprehensive, running through a series of tests in a predetermined sequence. On other systems it is necessary for the operator to call or load the programs and to decide on a logical order for the tests. No matter what the fault appears to be, always check the heart of the data system first; that is, run CPU and memory diagnostics before testing peripheral devices. This allows you to move logically from a known working part of the system to a suspected fault area.

Hardware Checks

Failure of a diagnostic does not automatically mean that a given unit is defective. This is especially true if interconnections are involved. Computer equipment often runs quite hot, especially in GC/MS cabinets standing in laboratories instead of computer rooms. They usually have fans to blow cooling air through them. Always check whether the fans are working, the power supplies are good, and connectors are clean and in good condition. Data system cables are prone to damage, especially the ribbon variety. Failure of a cable either as a short or open circuit can lead to all sorts of peculiar effects. Sometimes these can be seen as unusual signal ratios, such as strange isotope patterns or as bit dropouts, as shown in Figure 157. In this case, the problem was a faulty cable drive circuit on a 16-bit parallel interface. The drive circuit had blown due to an intermittent short on the cable, causing excessive loading.

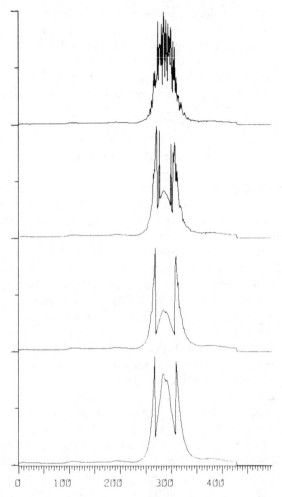

Figure 157. Effect of a bit dropout on a GC/MS data system interface during MID acquisition. In this case the fault was due to a cable short circuit. Four ions were monitored. The upper mass chromatogram just appears noisy, but something is obviously wrong with the other three ions.

Disk and Tape Units

These rely on precise registration between the read/write heads and the recording media. Random loss of data can be caused by alignment problems and dirt buildup on the recording head. Always follow the manufacturer's recommended procedure when cleaning tape and disk heads, or serious damage to both the drive and the recording media may occur. Alignment of disk drives and the condition of disks should be checked periodically, at least once a year. If a head crash, however slight, occurs, the disk enclosure will be contaminated with fine particles and should be thoroughly cleaned. This is a job for an expert. Never put a suspect disk into a disk drive unit or good disks onto a suspect drive. Shuffling the disks only compounds problems.

Tape units are not quite so critical as disk drives but should still be treated with respect. Alignment problems are often along, rather than across, the tape and can be caused by tape slip. Check the drive pulleys to make sure that their surface has not become glazed and slippery.

Pilot Error

The complexity of modern digital electronics compared with the ease with which they generate displays and compute results can lull one into a state of complacency. Do not forget that the data system can only follow prescribed procedures and that the responsibility for reliable data lies with the operator. A lot of computer faults are in fact operator errors, which can go undetected if results are not carefully reviewed. Many faults are wrongly attributed to computer bugs when the operator (or programmer) makes a mistake. For instance, one group of operators were using data from quite an old textbook with mass data based on ^{16}O instead of ^{12}C. Their data system computed nominal mass data and so this was not much of a problem; mass assignments were usually correct. Unfortunately, they did a lot a multiple-ion scanning work setting the selected mass to 0.1 u. Rarely did they hit the mass peak tops, and so their data were very unreliable!

PART FOUR

CHOOSING
A GC/MS SYSTEM

CHAPTER THIRTEEN

Instrument Selection
and Evaluation

This part of the book is potentially controversial. If you already have a GC/MS system, you may feel that you do not need to specify an instrument; like it or not, you have it. If you are about to purchase an instrument, or recommend the purchase of one, you will probably already be inundated with advice, a lot of it conflicting. The advice given here is not entirely unbiased. Manufacturers of GC/MS equipment may feel that insufficient details of their latest developments are included here. Unfortunately, it is difficult to be right up to date.

In my experience, most GC/MS laboratories have what they would call a satisfactory instrument, but there are some who have unsatisfactory instruments—instruments that do not do what the laboratory thought they would do or now wants them to do. Obviously, an unserviceable or unreliable instrument is unsatisfactory to any user. No one puts up with that for long, but many laboratories live with poor performance or unusable features because of specification and purchasing misunderstandings. Of those who have satisfactory instruments, a high proportion have adapted to their instrument's limitations. In other words, their actual requirements are now quite different from those conceived at the time of purchase. Many laboratories do not use certain parts of their instruments or make use of some of the features, even though they have paid considerable sums of money for them.

Some general comments will be of use to those considering the purchase of a GC/MS system and those about to request GC/MS analysis services. If you have an instrument already, a review of its capabilities and your initial and current expectations could lead to a more effective and successful use of the equipment. The manufacturers reading this will be aware that customers buy what they think they need, not necessarily what they actually need. For a sale to become a positive reference, these two needs, real and imagined, should be the same. If the instrument satisfies these, then the customer will be not only happy but enthusiastic about your product. Identifying his real needs, educating him if he is

329

unaware of these and then satisfying his real requirements must be a successful formula for all concerned.

DEFINING REQUIREMENTS

Usually when someone decides to buy a GC/MS system there is a perceived need or a problem that such an instrument may solve. It may be that the facility is already used but the existing instrumentation is unsatisfactory and a replacement is required. Perhaps an outside service or another laboratory currently handles the work, but it is felt desirable to bring the analysis of samples under more direct control by investing in the necessary equipment. In all cases, the laboratory's requirements and expectations should be assessed.

The Present Need

What Are the Immediate Requirements?   In many cases, a single factor may be the justification for the instrument. Will this factor still exist by the time the instrument arrives and the staff are trained to use it? If the answer to this question is no, then the use of a contract GC/MS laboratory would probably be a better solution. It does not make sense to buy a GC/MS system just in case a similar problem arises. For one thing, if the staff do not have enough work for the system, they are unlikely to become skilled enough to deal with the next crisis. On the other hand, if there is work for such an instrument on a regular basis, the proposition must be attractive.

What Type of Sample and How Many Must be Analyzed?   Try to divide the types of work into categories. Some categories will be incompatible in terms of the GC/MS system. For instance, pesticide residue work and the analysis of biological fluids using derivatizing reagents may require quite different analytical methods and it may not be practicable to switch quickly from one to the other. It is unlikely that you will buy two instruments initially to separate the two types of work, so some consideration must be given to workload planning. Will it be possible and/or necessary to process samples of different types in batches? The effect of batch processing on sample throughput and instrument availability must be considered.

The types of analysis required also dictates the data display format. If a lot of similar samples are to be analyzed, as in a clinical trial, then it is useful to have some way of handling and presenting the data as a set. Consider the normal mode of data presentation. Will data be viewed on a terminal by the originator of the sample, who will then select only the points of interest for permanent copy? If the operator is the main sample producer, then the instrument will be primarily dedicated to his or her work. In multiuser or multicustomer situations (the customer being the person requiring the analysis), a second display terminal may be desirable. Alternatively, if data have to be hardcopied in full and sent to the customer, then a high-speed plotting capability is important.

Data System Requirements. All this assumes that a data system is required. It is a natural assumption these days but not necessarily valid. The data system will account for about half the cost of a fully automated GC/MS system. Data systems allow all sorts of display and data-handling formats, but in some areas these are not required or at least not essential. The funds might be better spent on two separate manual instruments to handle the "chalk and cheese" samples separately, allowing greater efficiency and instrument utilization. This could be followed by a data system to handle both instruments, if and when it is required and can be economically justified. It is very easy to get carried away with the jargon and the processing power. One can forget that one is buying a GC/MS system possibly with computer control rather than a computer with a GC/MS instrument as a peripheral device.

If it is decided that a computer is required, then the requirement should be reviewed carefully. Is it a matter of speed, data handling and presentation, or data storage? Often it is a combination of all of these, under the umbrella term "laboratory efficiency." For a data system to be useful it must be easy to use. If it is as smart as they say, then the operator should not have to be. Routine should be just that—routine. All the usual and normal procedures should be straightforward, easy to use, and logical; logical to the GC/MS user, that is.

The required computer parameters should be defined, but you are probably not a computer expert and do not know all the buzz words. Define your requirements in your own terms. For example, you may be interested in five capillary runs a day, each of about one-hour duration. At one scan a second, that is 18,000 spectra per day. Working an 8-hour day, 3 hours are left for setting up and shutting down the instrument and preparing the samples. It would seem obvious that a lot of the sample processing must be done during sample acquisition. Let us suppose that each run typically contains 500 compounds, which must be survey searched against standards, and any other large unknown peaks must be library searched. You could soon produce mountains of paper, so data must be further reduced into a manageable form and stored as files on disk so that they can be quickly checked if the results warrant it. All in all, this is quite a task for the GC/MS data system combination.

Obviously, this would be impossible on a manual system, but that is no excuse if the data system cannot cope. This is your requirement, your reason for buying a data system. Can the system do it, if not why not? How close can it get? Could it handle the data on the basis of 2 s per scan, for instance? If the salesperson cannot answer these questions in terms that you can understand, either the system cannot cope or the salesperson cannot, in other words he or she does not understand your requirement. No amount of Fortran, graphics, bytes, or what have you features will help if it does not do what you need. Of course, what you ask may be impossible or unrealistic. Better to find out before rather than after purchase, especially if this is the basis for instrument funding.

Data Storage. Another area that seems to get out of proportion is data storage. Decide what you would like to keep and for how long, again in your own terms. For the example above, suppose you wished to keep some of your data forever, most of it for a few weeks, and all of it for one week. In other words, once a week (or more often) the data are to be reviewed and then stored or discarded as required. At first sight, you would need space for over 90,000 spectra and, of course, room to process them. In practice, by applying enhance routines, this number could be reduced initially by about a factor of 10. Enhance routines require a fair amount of working space, so some surplus capacity will be required. They also take time, and in our example you do not have much time to spare. Once again, try to assess your requirements and weigh them against the limitations imposed by your budget and the proposed system hardware configuration. A computerized system is hardly automatic if you must keep changing disk packs.

Tape units would seem a logical choice if you are obliged to store a lot of data, especially if the data are not viewed very often, if at all. A tape reel can store the equivalent of four disks and cost, say, £20, whereas a disk costs, say, £80, an apparent saving of £75 a disk.* However, if the tape transport costs about £6000 we need to fill the equivalent of 80 disks to break even, assuming all other costs are equal. There are, of course, other factors. Disk drives are not so reliable when disks are changed often. Disks require careful storage and occupy a lot of space. On the other hand, they are a fast storage device and can be read quickly. Tapes are much slower by comparison and the software does not always provide list and search facilities. In other words the data are dumped en bloc onto the tape and must be read back in the same way. The relative performance of different storage systems should be considered carefully.

Users.   Whatever the prices of the hardware and the pros and cons of the various options, it is up to the purchasers to decide what will best suit their needs. Sadly, the actual users of the system are not always consulted, the buying decision being made elsewhere in the organization. Where possible, the users should be actively involved in the choice of instrument and have a say in the selection process. Certainly consideration should be given to who will be operating the system.

How many operators will there be? Do they have the skills required to run and maintain the instrument, or will training be required? Have resources, funds and time, been allocated for operator training? Who will be in charge of the instrument on a day-to-day basis? These questions should be resolved before purchase, rather than after, so that individual responsibilities and levels of interest can be defined. Someone who is not going to be part of the GC/MS operator team should not exert a controlling influence over its choice. However, someone without a vested interest can often pick up points missed by those actively involved. The same advice applies to GC/MS customers. While they do not in principle care how data are obtained as long as they get the results from their samples, they do have valid comments to make. Consideration of their requirements must be a factor in the selection of a GC/MS system.

---

*The actual prices are liable to change but the figures give a fair indication of the relative tape versus disk costs.

Future Needs

Having established what is currently required of the GC/MS system, some thoughts should be given to its future use. Will the current workload continue? Does the instrument have the necessary features and facilities to handle a variety of analytical methods? If not, can these features be added? It is always a good idea to get quotations for extra features, giving the cost of adding them as a retrofit option. The quotations will be guarded, because materials and labor rates change, but some idea of the relative cost of adding an option at a later date should be obtained. Establish the likelihood of future availability of the up-grade features. It is also worth finding out if the option is really available now. In other words, has anyone else had the optional addition fitted after initial instrument delivery? How long did the conversion/addition take to fit, and how much instrument time was lost because of the retrofit work?

Quite apart from availability, will additional features really be viable? Consider the workload plan for the instrument. Will there be enough time for workload expansion, using the planned techniques and methods, let alone any upgraded facility? Many laboratories achieve less than 10% instrument downtime for whatever cause. Others are less fortunate and cannot improve on 50%. Some of this may be due to laboratory efficiency, mixed types of analysis and so on, as well as instrument operational factors. If your budget calls for only 50% instrument utilization, there is room for expansion. If the plan relies on less than 10% downtime, there is very little room for improvement. Certainly in the first year, it would be wise to aim for a more modest percentage; 75% system availability, 25% downtime is probably realistic. Where utilization is high and downtime low, the addition of facilities at a later date would not improve efficiency unless some samples were transferred from an old method to a new one using the new facility. For example, adding chemical ionization may increase the workload because new types of analysis may be undertaken, or it may improve efficiency because some samples will now be run more successfully under CI conditions instead of in the EI mode.

Performance

All new GC/MS systems will, or at least should, work at their stated level of performance. Operation at these levels may be somewhat artificial. Decide what level of performance is necessary for your present and future workloads and choose an instrument with sufficient power. In routine use, it is unusual for laboratories to maintain their instruments at the peak of performance. Some do; in fact, a few laboratories routinely achieve even better results than the published specifications, but they are the exceptions. Most laboratories do not push so hard or expect so much from their instruments. However, if it is not there it cannot be obtained when needed, so it is important to give careful consideration to performance figures.

The first thing to decide is, what is important in your circumstances? Are you most interested in sensitivity, scan speed, resolution, or what? Be realistic; you cannot have the best of everything. Decide what is critical and compromise only on the other factors. Beware of the numbers game. Do not just go for high value parameters because they sound good. Will a resolution of 2000 be any more use to you than a figure of 1000 or even 500 when looking in the low mass range, for instance 100 to 200 u? If you can think of a good reason for saying yes, then this is a valid specification for your instrument, but remember that bigger is not necessarily better.

There is often a tradeoff between one parameter and another. An instrument may have extremely impressive sensitivity and resolution figures that are mutually exclusive. In other words, it may be able to give high sensitivity or high resolving power but not both at the same time. Sensitivity and detection limit is another area where one should be careful. Sensitivity is often quoted for methyl stearate which is a known compound. Often, at the limit of sensitivity as defined in the brochures there is no discernible chromatographic peak. Obviously, when doing search work in either the full-scan or MID mode you know what you are looking for and high sensitivities can be achieved. On the other hand, the minimum detectable quantity of unknown substance, in a practical sense, is one that will give an observable chromatographic peak. Manufacturers of GC/MS systems are prepared to quote figures for spectral sensitivity and even the chromatography conditions such as retention times, but rarely do they quote signal-to-noise ratios for the chromatography trace; in other words, the detection limit for an unknown substance is not defined.

Budgets

The cost of running a GC/MS system is not insignificant. While there may be valid reasons for purchasing an instrument and money available to cover the initial cost, there must also be sufficient funds available to run the instrument on a day-to-day basis. Allowance must be made for wear and tear; some parts of the instrument are very expensive and yet have only a limited lifetime. Money must be put aside for their future replacement. A £1000 multiplier is an expensive item. It may last 2 or 3 years. Similarly, items such as jet separators and parts of the ionizer may be handled quite often, and although they do not wear out as such they are relatively fragile and are easily damaged. A fund should be established to cover such catastrophes or, better still, it is advisable to purchase and have on hand a few vital spare parts.

GC/MS instruments also consume on an on-going basis quite a lot of expensive items. Things such as paper, ferrules, and septa are obvious examples, but there are also the less obvious, like computer disks and tapes on automated systems, which are consumed by virtue of the fact that they are filled with data. Disks and tapes can be reused only if the data are overwritten and therefore lost. Not quite consumables in the above sense, but requiring regular replacement nevertheless, are maintenance items. Pump oils and O-rings for the vacuum system, cleaning materials for the mass spectrometer, and filters for the data system are typical examples.

To run a GC/MS laboratory also requires skilled operators and support staff. The cost of running the facility must include a budgetary allowance for staff of the required caliber. Training costs, both in the initial stages when the laboratory is first set up and later when new staff come into the GC/MS group, must also be considered. Many laboratories limit their staffing requirements by subcontracting some of the maintenance and service work. Service contracts come in a number of different forms and the cost depends on the level of support offered. Although a service contract may not be necessary during the initial warranty period, it is worth getting typical estimates along with equipment purchase price quotations. This will enable you to assess the cost of maintaining the instrumentation out of warranty.

All in all, the running costs of a GC/MS facility are considerable and should be carefully considered before setting up the laboratory or purchasing new and more sophisticated equipment. Any justification for an instrument purchase must consider all the running costs as well as the initial purchase price.

EVALUATION OF A GC/MS SYSTEM

Having decided that a GC/MS system is required and that funds are or at least may be available, you must decide which instrument to buy. More than this, you have to decide which options and features are needed to satisfy the defined requirements. Manufacturers' brochures and specifications are only a guide to an instrument's true performance. Also, performance is not the only criterion; the instrument should be easy to work with, it should feel right. This is a very personal thing. There is no best instrument. They all have both good and bad features, and the good become the bad and the bad the good under different circumstances.

Test the instruments with your particular type of work. Send some samples for analysis, and examine the results produced. Arrange for a demonstration. Get a feel for the instrument, the way it handles, and the organization that will be supporting you if you buy it. Gen an idea of its running costs and the maintenance work required to keep it in top form. Gather opinion from other users, visiting them if possible, and review the areas where a particular instrument has been successful or otherwise. This may reveal interesting aspects of a given instrument's capabilities and practical limitations.

Test Samples

Long before visiting a GC/MS manufacturer's demonstration facility, probably even before you have actually decided that GC/MS is the solution to your problem, you will need to know if it can handle one of your own samples satisfactorily. Remember you are trying to evaluate an instrument and possibly the technique. You are not trying to test the operator, after all when you get your instrument you will have the best operator—yourself or someone you appoint.

Neither are you trying to complete your part—time thesis by analysing a variety of speculative unknown samples on the cheap.

To evaluate an instrument you need to test it with a known problem. In other words, you must already have the answers. To do this, you must use a prepared standard sample, be it a single compound or a complex mixture. Use a variety of samples if necessary, but not too many, as this is unfair to the applications laboratory. You do not want a rush job on a large number of samples but the best results that are reasonably possible on a few. Stop for a moment and imagine you have bought the instrument and you are running it for one of your customers. It is very unlikely that the customer would not supply any information about the sample. In fact, many laboratories will not run samples unless a gas chromatogram with conditions is first supplied. GC/MS time is at premium rates. Do not test the skill of the manufacturer's applications chemists; give them as much data as possible when you submit the samples. Having done this you can reasonably expect a quick turnaround with viable data.

Of course, data can be "fudged," and you would be wise to be at least a little cautious. With your well-defined standard test samples, send one or two "unknowns." These should have identical chromatographic conditions but be of unstated concentration, perhaps spiked with an "unknown" compound. At this stage, you need not alter the parameters very much, just enough to remove the temptation to cheat.

Identical samples should be sent to all manufacturers whose instruments you are interested in. This is the only way to make realistic comparisons. If you wish to test a particular feature on a given instrument, send an additional sample to test that area. Be specific about the expected results and the data required. This saves having to wade through a ream of computer printout with an overzealous salesperson, when all you want to know was whether the system has sufficient scan speed and chromatographic resolution to handle 20 peaks a minute on a 20-minute capillary run.

The purpose of sending samples for analysis is to establish that the instrument has the necessary performance and features to tackle your type of problems. If they make a hash of the analysis, do not write the instrument off. Give them a second chance. There are not so many instruments to choose between that you can afford to let one go because of a human error. Test the instrument, not the people (at this stage anyway). It is also just possible that you made a mistake with the samples or that they were degraded in transit. The more complicated a sample or its preparation, the more likely this may be the case. Run a gas chromatography trace yourself before sending off the samples to save yourself from embarrassment later if you have made a mistake.

Demonstrations

The pace is quickening. You have decided to visit a few of your candidates. If the initial sample results were not satisfactory, do not go for a demonstration. If they gave poor results without a reasonable explanation on two lots of samples (providing you followed the foregoing advice), then the instrument will not do. Do not waste your time and theirs. If you go for a demonstration in these circumstances, you will start comparing apples with golf balls or something (even the good instruments are going to have apple/orange comparisons, so do not complicate things.) You will only confuse matters if you go and will find it harder to say "No" in the end.

Once again, prepare known samples and supply relevant information before visiting the manufacturer. If possible, keep to the same conditions as the earlier samples you submitted. This will ensure minimum time for setting up and sample preparation. You probably know about the importance of using the right phase in a well-conditioned column; you do not need to have it demonstrated. You need to assess the instrument and how it is used. Sticking to a simple or repeat sample to start with allows you to concentrate on the instrument's features. Remember that no matter how good the instrument and how easily analyzed the sample, the manufacturer-customer interchange is essentially one of personal communication. Give the demonstrator a chance to get to know you. The demonstrator has a vested interest in showing the instrument at its best, and you have a vested interest in seeing it at its best. Do not present an extremely complex problem to be solved in double-quick time. Similarly, do not be blinded by science. Especially beware of elaborate canned demonstrations of features that you are not interested in. An overview of the whole system is reasonable, essential in fact, but should not exclude your own areas of special interest. Do your homework and present your questions openly and clearly. This will encourage a similar open response.

Assuming all goes well, you will soon be able to move into unknown territory. Up to now all the samples have been known, even if all the data have not been revealed. The time now comes to try out the real problems. Obviously, if the required conditions are the same or similar to the previous samples, then it is a smooth process to move on. If you want to see packed versus capillary work, EI versus CI, high and low resolution, or other contrasting features, treat them as separate parts of the demonstration. Try to plan things in a sensible sequence. If you are unsure of the best order, advise the demonstration laboratory of your requirements in advance and let them choose.

Having come to the true unknown or difficult sample, you should be in a position to evaluate the instrument's potential for solving your problems. A positive result is always satisfying for all concerned but is not necessarily the main criterion of success. Demonstrations are at best limited. There is always a great deal to see and do and only a short time available. Try to get a feel for the way the instrument can be used to obtain a solution to your problem and how easy and practical it is to do so. If time and circumstances prevent you from getting a satisfactory result, be prepared to leave the sample for further work by the applications chemist. By this time, you should be able to assess his or her ability and the basic instrument performance and be in a position to judge the consequent data realistically.

What If the Demonstration Goes Badly?   We all have our off-days, so do not be too hasty in condemning the instrument or the people. On the other hand, demonstrations take a lot of time, and you must consider carefully whether it is worthwhile to go back. Leaving some of the samples for further analysis by the demonstration laboratory is a reasonable compromise, allowing you time to consider the situation. Even if the day goes very badly, you should be able to assess the general hardware of the instrument and decide if you like its features. This stressful situation also gives you a good opportunity to evaluate the organization and will afford more information than you would obtain on a good day. If the applications chemist cannot get good support, what chance will the potential customer have? The answer is not always as bad as it would seem. Applications laboratory instruments are often used as test instruments to support customer service. Also they are operated in a multi-user, varied sample environment and consequently are not always at their best. You must consider the circumstances and judge for yourself.

After-Sales Service.   When you visit an organization for a demonstration, do not be content with seeing just the applications laboratory and its staff. Look around the rest of the outfit. Speak to the support staff. Once you have bought the instrument, these are your main points of contact within the company. Find out what facilities and services are available to support you and your instrument and who is responsible for these functions. For instance, who will be responsible for installation, and how will it be organized? What training will be given? In addition to basic training when the instrument is delivered (if any), what on-going training facilities and arrangements are there? How will the instrument be supported? What are the service stocks of spares like? How many field engineers are there in your area? Find out about the types of service contracts offered. Often companies will charge a premium if a contract is taken out some time after the end of the warranty period, so it is a good idea to get details early.

Get an idea of the level of spares backup. Does the company hold stocks of the major consumable items as well as any critical components, such as multipliers and separators? In some parts of the world, companies operate special stores arrangements or handle spare part and consumable sales through dealers and distributors. Much of this kind of detail about the people and the organization can be obtained during a demonstration visit.

Data System Considerations

The developments in the electronics field, particularly in the area of computing, have been so rapid in recent years that even those working in this field have found it hard to keep up. As a result, anyone buying a new piece of scientific equipment has been faced with quite a dilemma. Obviously a computer or some sort of automatic control and data processing system would be the right thing to have as part of the proposed GC/MS system, but what features should it have? The computer layman will find this question hard to answer. What is more, for many people any form of computer is exciting, confusing, or both. Without doubt, you must assess the various data system options or automation packages very carefully. As it is so easy to be confused with all the buzz words, try to nail down exactly what they mean and what relevance they have when discussing the features with the sales and demonstration staff.

At a demonstration, try to determine how easy it would be to use the data system. If possible, operate it yourself. You will have to be guided by the demonstrator, but you should be able to gauge how easy it will be to learn to drive the system yourself. Do not be put off by the power of the system or by the range of available options and commands. All these features are there to cope with a wide variety of circumstances. Most of the time you will be using only a few basic commands and routines. Do these seem logical and straightforward? The things you do most often, such as viewing a portion of an RIC trace and displaying spectra, should require only few keystrokes. Comprehension of the more exotic routines and the skill to use them will come with use and training. If the basics are logical and easy to understand, the chances are that the rest will not be too difficult.

Up to a point, you should not be too bothered about the data system specification. Its operating language, the amount of storage, the number of processors, and so on are not relevant so long as the system does the job. If Company A gives you two computers working side by side in one box while Company B offers one computer working twice as fast, what does it matter? If you can get your data, control your instrument, and store your results, then all is well; if not, the data system is not satisfactory. Certainly the way the data system hardware and software are implemented will affect its performance. Remember that you are buying a GC/MS system, not a computer. If a certain feature is claimed to offer a significant advantage, ask for it to be demonstrated so that you can see first-hand how it improves the GC/MS system. Then you can judge for yourself how important it is. When you visit other companies you can pose a similar problem to see how they manage without this wonderful feature. There may be a number of ways of achieving the same result.

On the other hand, some features are fundamental. Can data really be processed (a background job) during acquisition (the foreground job) on systems with a foreground/background capability. The limitations of the systems offered should be discussed. Can enhancements or print routines be done efficiently during high data rate acquisitions? How many terminals can operate on the system at the same time? Try to read between the lines of the sales brochures so that you can assess potential limitations.

Another critical area is the software. As suggested earlier, examine the flexibility and ease of use of the system. How user-friendly is it? Does it give you prompts that lead you to the correct commands, or do you have to remember elaborate procedures? If you forget a command, how easy is it to find? Does the system have a query file to prompt you, or do you have to look it up in the manual? How good is the documentation? It may look impressive, but is it useful? During the course of a demonstration many routines will be shown. Ask for one to be repeated in slow motion, so that you can note the sequence of events and commands entered. Than ask to see the relevant documentation to see if it would lead you through the procedure adequately.

Many computer manuals (and other manuals, for that matter) are written by people who already know the answers, and they tend to be very brief, almost reminders or check lists. This is fine when you know the system well but very frustrating to a beginner. Some companies offer a beginner's manual that has most of the basics in some detail so that you can quickly get to grips with the system. Adequate on-site training during installation and afterwards is also vital. Establish at an early stage what help comes as standard with the instrument and what other courses are available.

MAKING UP YOUR MIND

If a demonstration goes well, the potential customer is often left feeling a little uneasy in case it just happened to be a very lucky day. Bad days are usually

Figure 158. Some of the finer points of instrument demonstrations. (Reproduced with permission from *European Spectroscopy News*, Vol. 16, 1978)

considered to be normal. The same is true of existing customer opinion. Good reports are treated with skepticism (who's paying him?), while bad words are believed without question. Just as you should be analytical in your GC/MS work, so too should you be in your assessment of opinion. If possible, contact a number of users. Have ready a few particular questions about the instrument and their work with it as well as more general questions. Opinion will depend on the day and the pressures of work at the time of the call. Specific questions about the work, sample throughput, and performance will tend to get specific answers, whereas general questions such as "What do you think to the instrument?" will depend on the person's mood at the time. Where possible, visit a laboratory to gather opinion and experience of the instrument. This will allow you to judge the opinion in the light of the experience of the people and their working environment.

Finally, when you purchase your instrument be sure the purchase agreement clearly states what you are getting. Things are often promised or implied but not defined in the contract. The installation engineer may not know he has to undertake some extra work or demonstrate a special specification. A lot of goodwill can be lost and bad feelings caused just at this crucial point when a working, friendly relationship between customer and manufacturer should be established. If you have special requirements, get them in writing, not just in the quotations but also in your letter of acceptance. Do not forget to read the small print of the contracts and guarantees. The instrument will be your responsibility. It is up to you to get the best out of it.

Throughout this section, I have referred to *instrument features*. I once went to a lecture where this was the euphemism for a bad point or useless parameter. I have tried to be fair throughout, but I realize I am biased. In fact, we are all biased, and it is well to remember this when choosing an instrument. You will have to live with your choice for some time, so try to assess the features carefully. I am indebted to the journal *European Spectroscopy News* for permission to publish the cartoon in Figure 158, which shows some of the finer points of instrument demonstrations and evaluations.

Further Reading

The following is a list of books, papers, and articles I have found useful in my GC/MS work and in the preparation of this book. A search in a good library will yield many more. The few works I have quoted directly are detailed in the text at the point of interest.

Budde, W. L., and J. W. Eichelberger, *Organics Analysis using Gas Chromatography/Mass Spectrometry—A Techniques and Procedures Manual*, Ann Arbor Science.

Eichelberger, J. W., L. E. Harris, and W. L. Budde, "Reference Compound to Calibrate Ion Abundance Measurements in Gas Chromatography-Mass Spectrometry Systems," *Analytical Chemistry*, Vol. 47, No. 7, 1975, pp. 995–1000.

Ettre, L. S., "Introduction to Open Tubular Columns," GCD-46, Perkin-Elmer Corporation, Norwalk, CT, 1979.

Feser, K., and W. Kögler, "The Quadrupole Mass Filter for GC/MS Applications," *Journal of Chromatographic Science*, Vol. 17, 1979, pp. 57–63.

Grob, K., and K. Grob, Jr., "Splitless Injection and the Solvent Effect. *Journal of High Resolution Chromatography and Chromatography Communications*, July 1978, pp. 57–64.

Hunt, D. F., "Selective Reagents for Chemical Ionisation Mass Spectrometry," *Finnigan Spectra*, Vol. 6, No. 1, February 1976.

Hunt, D. F., G. C. Stafford, F. W. Crow, and J. W. Russell, "Pulsed Positive Negative Ion Chemical Ionisation Mass Spectrometry," *Analytical Chemistry*, Vol. 48, No. 14, 1976.

Kiser, R. W., *Introduction to Mass Spectrometry and its Applications*, Prentice-Hall, Englewood Cliffs, NJ.

McFadden, W. H., *Techniques of Combined Gas Chromatography/Mass Spectrometry*, Wiley-Interscience, New York, 1973.

McNair, H. M., and E. J. Bonelli, *Basic Gas Chromatography*, Varian Aerograph, 1969.

Michnowicz, J. A., "Reagent Gas Selection in Chemical Ionisation, Mass Spectrometry," Hewlett Packard Application Note AN176-13.

Millington, D. S., "New Mass Spectral Techniques for Organic and Biomedical Analysis: A Review," VG Organic Ltd., 1980.

Oswald, E. O., P. W. Albro, and J. D. McKinney, "Utilisation of Gas-Liquid Chromatography coupled with Chemical Ionisation and Electron Impact Mass Spectrometry for the Investigation of Potentially Hazardous Environmental Agents and the Metabolites," *Journal of Chromatography*, Vol. 98, 1974, pp. 363-488.

Rapp, U., G. Meyerhoff, and G. Dielman, "DCI-Direct Chemical Ionisation or Desorption Chemical Ionisation. A powerful soft ionisation technique in mass spectrometry," Finnigan MAT Application Report No. 47.

Slayback, J. R. B., and M. N. Kan, "Biomedical Applications of Negative Ion Chemical Ionisation Mass Spectrometry," Finnigan MAT Application Report No. 8020.

Steinherz, H. A., and P. A. Redhead, "Ultrahigh Vacuum," *Scientific American* March 1962.

Taylor, D. M., "Mass Spectra of Common GC Liquid Phases," *Finnigan Application Tips*, No. 54.

Warren, P. L., J. R. B. Slayback, and C. R. Phillips, "Optimisation of Acquisition Parameters for the Incos Data System," Finnigan MAT Technical Report No. TR8027.

Yost, R. A., and C. G. Enke, "Triple Quadrupole Mass Spectrometry for Direct Mixture Analysis and Structure Elucidation," *Analytical Chemistry*, Vol. 51, No. 12, 1979, pp. 1251-1264.

INDEX